Vesselin Petkov (ed.)

I0030970

THE ORIGIN OF SPACETIME PHYSICS

With a Foreword by Abhay Ashtekar

MINKOWSKI
Institute Press

Minkowski Institute Press
Published 2020

ISBN 978-1-927763-70-4 (softcover)
ISBN 978-1-927763-71-1 (ebook)

Minkowski Institute Press
Montreal, Quebec, Canada
minkowskiinstitute.org/mip/

For information on all Minkowski Institute Press publications visit our website at
minkowskiinstitute.org/mip/books/

PREFACE

This volume contains 13 papers by Lorentz, Poincaré, Einstein, Minkowski, de Sitter, Friedmann and Lemaître that laid the foundations of spacetime physics and a very informative Foreword by Abhay Ashtekar.

Seven of the papers (two by Lorentz and five by Einstein) were first published together in 1923 in the collection[1] *Das Relativitätsprinzip: Eine Sammlung von Abhandlungen*[2] which was translated into English and published the same year as *The Principle of Relativity: A Collection of Original Memoirs on the Special and General Theory of Relativity.*[3]

In addition to the seven papers, the present collection includes new translations from French (by André Michaud and Fritz Lewertoff) of Poincaré's two papers "On the Dynamics of the Electron" (dated 5 June 1905 and 23 July 1905), a new translation[4] from German (by Vesselin Petkov) of Minkowski's paper "Space and Time," a new translation[5] from Russian (by Vesselin Petkov) of Friedmann's paper "On the Curvature of Space," de Sitter's paper "On the Curvature of Space" and Lemaître's paper "A Homogeneous Universe of Constant Mass and Increasing Radius accounting for the Radial Velocity of Extra–Galactic Nebulæ."

The purpose of the present collection is to make these papers easily available[6] together to all interested in the origin of spacetime physics. All papers

---

[1]The first English translation of Minkowski's paper "Space and Time" was also in that collection; here a new translation is included.

[2]H. A. Lorentz, A. Einstein, H. Minkowski, *Das Relativitätsprinzip: Eine Sammlung von Abhandlungen.* Mit einem Beitrag von H. Weyl und Anmerkungen von A. Sommerfeld. Vorwort von O. Blumenthal, Fünfte Auflage (Springer Fachmedien Wiesbaden Gmbh, 1923).

[3]H. A. Lorentz, A. Einstein, H. Minkowski and H. Weyl, *The Principle of Relativity: A Collection of Original Memoirs on the Special and General Theory of Relativity.* With Notes by A. Sommerfeld. Translated by W. Perrett and G. B. Jeffery (Methuen and Company, Ltd., 1923; reprinted by Dover Publications Inc., 1952).

[4]First published in Hermann Minkowski, *Spacetime: Minkowski's Papers on Spacetime Physics* (Minkowski Institute Press, Montreal 2020), Translated from German by Gregorie Dupuis-Mc Donald, Fritz Lewertoff and Vesselin Petkov. Edited with an Introduction by Vesselin Petkov.

[5]First published in Alexander A. Friedmann, *Papers On Curved Spaces and Cosmology* (Minkowski Institute Press, Montreal 2014), Translated from Russian and edited with an Introduction by Vesselin Petkov.

[6]Finding some of the papers in this collections is not always easy – for example, it seems an electronic version of de Sitter's paper is available (as a PDF file of a very poor resolution) only at the Digital Web Centre for the History of Science (associated with the Huygens Institute and the Royal Netherlands Academy of Arts and Sciences) – `https://www.dwc.knaw.nl/DL/publications/PU00012216.pdf`.

i

- were proofread

- all noticed errors and typos (especially in the equations) were corrected

- were typeset into LaTeX (which is another reason for the publication of this collection).

I would like to thank Abhay Ashtekar for the excellent and informative Foreword, which makes this collection even more valuable.

I would also like to thank André Michaud and Fritz Lewertoff for the translation (as always on a voluntary basis) of (i) Poincaré's two papers and (ii) the first page of Einstein's 1916 paper "The Foundation of the General Theory of Relativity" which was missing in both 1923 publications of *Das Relativitätsprinzip* and *The Principle of Relativity*.

I would like to express my gratitude to my wife Svetla Petkova for the hard work of typesetting 12 of the 13 papers into LaTeX and to Reinoud Slagter for typesetting Poincaré's longer paper "On the Dynamics of the Electron" (dated 23 July 1905) into LaTeX.

Montreal
28 December 2020

Vesselin Petkov
Minkowski Institute

FOREWORD

Let us begin with a summary of how the notions of space and time have changed over the centuries, making full use of hindsight and using contemporary terminology. For nearly two thousand years, the four books Aristotle wrote on physics provided the foundation for natural sciences in the western world. In modern terms one can say that in Aristotle's paradigm, there was absolute time, absolute space, *and* an absolute rest frame, provided by earth. This was the reigning world-view Isaac Newton was exposed to as a student at Cambridge in the years 1661-65. Some twenty years later, Newton toppled this centuries old dogma. Through his Principia, first published in 1687, he provided a new paradigm. Time was sill represented by a 1-dimensional continuum and was absolute, the same for all observers. All simultaneous events continued to constitute 3-dimensional space. *But there was no absolute rest frame.* Thanks to the lessons learned from Copernicus, earth was removed from its hitherto privileged status. Galilean relativity was made mathematically precise and all 'inertial' observers —moving uniformly with respect to one another— were put on the same physical footing. Already in the beginning of the 1700s, papers began to appear in the Proceedings of the Royal Society, predicting not only the motion of Jupiter but even of its moons! No wonder, then, that the Principia became the new orthodoxy. It reigned supreme in what was then called *natural philosophy* for over a century and a half.

The first challenge to the Newtonian world view came from totally unexpected quarters: advances in the understanding of elecromagnetic phenomena. As is well known, in the middle of the 19th century James Clarke Maxwell achieved a grand synthesis of all the then accumulated knowledge concerning these phenomena in just four vectorial equations. Surprisingly, these equations further provided a specific value c of the speed of light that did not refer to a reference frame. But the notion of an absolute speed blatantly contradicted Galilean relativity, a cornerstone on which the Newtonian model of space-time rested. Since the model had been so enormously successful, the scientific community found it natural to conclude that Maxwell's value c of the speed of light is relative to a physical, privileged frame (even though the derivation itself made no reference to it). This was identified as the rest frame of *ether*, a medium permeating space that carries electromagnetic waves. But by doing so, the community in fact reverted back to the Aristotelian view that Nature specifies an absolute rest frame –but now anchored with ether rather than earth.

The first part, *Emergence of the Spacetime Continuum,* of this volume starts at this juncture. It comprises of seven papers by Hendrik Lorentz, Henri Poincaré, Albert Einstein and Hermann Minkowski. By the beginning of the 20th century a series of experiments had been performed with increasing sensitivity, the most influential of which was carried out by Albert Michelson and Edward Morley. The underlying idea was to measure the velocity of earth relative to ether –the so called 'ether wind'– by analyzing the speed of light in different directions and at different times as earth spins and moves around the sun. However, as Michelson wrote to Lord Rayleigh in 1887, *"The Experiments on the relative motion of the earth and ether have been completed and the result decidedly negative."* This null result was confirmed by other even more sensitive experiments carried out between 1902 and 1904.[7]

The first four papers by Lorentz and Poincaré illustrate that leading scientists at the time were sufficiently troubled by the ensuing dilemma as to propose radical new ideas. It is important to note that they both continued to use the notion of ether as a basic ingredient in their reasoning. Poincaré, for example, argued as follows: *"Does our aether actually exist ? ... If light takes several years to reach us from a distant star, it is no longer on the star, nor is it on the earth. It must be somewhere, and supported, so to speak, by some material agency."* Their analysis was based on Maxwell's equations, Lorentz's electron theory, and a shift from Augustin-Jean Fresnel's 'mechanical ether' to an 'electromagnetic ether,' whose states were now characterized by the electric and magnetic fields. The goal was to construct a theory which explains the null results of the 'ether wind' experiments. Their way out of the quandary can be understood as the introduction of length contraction and time dilation for all phenomena, when described in the reference frame of ether. The result is what we call *Lorentz transformations*, although one should note that the proof of the desired covariance of electrodynamics was established by Poincaré, who corrected the transformation properties of charge and current density that Lorentz had originally used and also used group properties of the full set.

The true paradigm shift came in the celebrated 1905 papers of Einstein's that follow Lorentz's an Poincaré's. Right in the beginning of the first paper, Einstein made a clear, decisive statement: *"The introduction of a "luminiferous ether" will prove to be superfluous in as much as the view here to be developed will not require an absolutely stationary space ... "* This insight is a masterpiece of the conceptual break-through that accompanies the removal of 'excess baggage'! Note that even though Poincaré presented the cleanest treatment of Lorentz transformations, he seems to have held the view that clocks resting in the aether show 'the true time', and even as late as 1912, he described light as the *"luminous vibrations of the aether"*. In a single stroke Einstein made Lorentz transformations kinematical statements about space-time structure, freeing them from 'properties of matter relative to ether'. In doing so, he not only abolished the absolute rest frame à la Aristotle that had crept back in through ether, but also the Newtonian notion of absolute simultaneity. Space-time *could still be* foliated by preferred 3-dimensional spatial sections of simultaneous events but the choice is no longer canonical – there is a 3-parameter family of these

---

[7]A century later, experiments using optical resonator have improved the accuracy of the null result and tested the local Lorentz invariance to one part in 10^{17}! See for example Ch. Eisele, A. Yu. Nevsky, and S. Schille, Phys. Rev. Lett. **103**, 090401 (2009).

foliations, one for each choice of inertial frames that are in uniform motion relative to one another.

The last two papers in the first part are by Minkowski. They complete this revolution by genuinely fusing together space and time into a 4-dimensional continuum of events. The role played by the *spacetime* metric comes to the forefront. In Newtonian theory, given any event, there is a unique 3-dimensional 'space section' of events that are simultaneous with it. This absolute space of Newton's is now replaced by a 2-sheeted cone that separates the region which is causally connected with the given event, from the region which is not. These light cones also dictate the causal propagation of fields. In the limit in which the velocity of light tends to infinity, the light cone becomes more and more 'wide-angled' and, in the limit the two sheets merge and morph into Newton's absolute space. The Poincaré group is now simply the group of spacetime transformations that preserves the Minkowski metric. Space-time geometry provided by the metric makes the underlying physics transparent. This is the conceptual framework that continues to underlie all of non-gravitational physics even today, including the quantum theory of fields. And it served as the key pointer in Einstein's efforts that led to general relativity.

The second part, *Gravity is but Geometry,* of the collection contains two papers by Einstein that serve as milestones in this transition. In 1907 while writing a review of special relativity, Einstein realized that while special relativity had successfully reconciled the predictions of Maxwell's electrodynamics, it was in deep conflict with Newton's theory of gravity. For, Newton's law of universal gravitational attraction required absolute time: the gravitational force between two bodies is transmitted instantaneously. Special relativity, on the other hand, had abolished absolute time. So Newton's law can not even be stated covariantly in Minkowski spacetime. In 1907 then, Einstein set himself the task of finding a new, better theory of gravity by generalizing the Minkowski geometry. The first paper included in this collection was written in 1911 and focuses on the influence of gravity on the propagation of light. It concludes that light rays from distant stars, passing close to the sun, are deflected by its gravitational field and works out the deflection to first approximation. Previous calculations based on Newtonian gravity had used the corpuscular theory of light. Einstein used the equivalence principle, and null rays of a metric, that is modified from Minkowski's in the 't-t' part by the gravitational potential of the sun. But the final result reported in this paper is only 1/2 the correct value (i.e. 0.87 rather than 1.75 arcsec), because it ignored the fact that the 'spatial part' of the metric is not flat in this approximation. The value he obtained was the same as that from the corpuscular theory of light in Newtonian gravity. In 1912 an Argentinian team was scheduled to measure this light deflection in Brazil and, in 1914, a Russian team planned to carry out the measurement in Crimea, both during solar eclipses. The first mission encountered bad weather and the First World War intervened the second. Had they been fully successful, they would have found that Einstein's result was incorrect and then Einstein's views on incorporating gravity in special relativity by modifying the metric would have suffered a major setback. So it is extremely fortunate that Einstein had more time to re-examine the calculation and obtain the correct expression well before Arthur Eddington led the successful mission in1919! The second, 1915, paper by Einstein in this part of the collection is the culmination of his 8 years of work. In contrast to Newton's *Principia*, gravity is no longer a force between two bodies, acting at a distance on one another. Rather, gravity is now encoded in the very geometry of spacetime. Massive bodies like sun bend the spacetime continuum of Minkowski's and light bodies like planets move in this curved spacetime, following the geodesic motion to first approximation. This is perhaps the most profound paradigm shift that physics has witnessed. As Einstein wrote to Arnold Sommerfeld, two days after he sent his paper to the Prussian Academy, *"During the last month, I experienced one of the most exciting and most exacting times of my life, true enough also one of the most successful."* The nature of the outcome is succinctly summarized by Hermann Weyl: *"It is as if a wall which separated us from the truth has collapsed. Wider expanses and greater depths are now exposed to the searching eye of knowledge, regions of which we had not even a pre-sentiment."*

The third part, *Birth of a Dynamical Universe* contains five papers by Einstein, Willem de Sitter, Alexander Friedmann and Georges Lemaître on cosmology. Recall that Newton made a profoundly bold leap and declared that the law of gravitation is universal – the same law that governs the fall of the apple on earth also governs the orbits of planets around sun. Einstein made an even bolder assertion. Equations he derived in order to bring gravitation in harmony with special relativity had received direct support only through small corrections to the orbit of mercury. But he made a giant leap and declared that they also dictate the large scale dynamics of the universe as a whole! This is the birth of modern cosmology.

At the time, the lesson Copernicus taught us was well appreciated and it was natural for Einstein to suppose that, on a sufficiently large scale, there is no preferred place, nor a preferred direction. However, one belief from ancient times still persisted: while change was manifest in our immediate neighborhood, on larger scales the universe as a whole was serene, distant stars unchanging and fixed, havens immutable. Therefore in 1917 Einstein sought a *static*, spatially homogeneous and isotropic solution of his field equations, assuming that the universe is spatially closed, and stars and nebulae can be modeled as dust particles. He realized that the requirement of staticity could not be satisfied if one uses his originally proposed field equations and therefore, somewhat reluctantly, introduced a positive cosmological constant λ. The result is the Einstein's cylindrical universe with topology $S^3 \times R$. In 1918, de Sitter showed that Einstein's equations with a positive λ admit other natural solutions with maximal symmetry in which there is no matter. His view was the presence of the physical matter in the universe contributes very little to the curvature compared to λ and could therefore be neglected in the leading order approximation. The topology is again $S^3 \times R$ but there is no matter; curvature is supported entirely by the positive cosmological constant. de Sitter presented this solution *as static*, conforming to the widely held belief in the unchanging havens (by using certain coordinates that cover only a part of what we now call the global deSitter solution). The solution drew attention because it naturally led to a redshift that seemed compatible with observations of extragalactic nebulae.

In 1922 Friedmann wrote two novel papers, deriving an entirely different class of solutions with matter, and allowing a cosmological constant of either sign. The first followed Einstein and de Sitter and considered closed universes . These solutions were independently rediscovered by Lemaître in 1927, who also allowed non-zero pressure, arguing that it cannot be neglected for radiation. The second Friedmann paper analyzed the 'open topology' in detail. In 1925 Lemaître clarified that the widely held belief that de Sitter universe is static was incorrect: apparent staticity was an artifact of the choice of coordinates in which spatial sections fail to be homogeneous and isotropic. Once homogeneity and isotropy are made manifest, the solution resembles the Einstein's cylindrical universe where, however, *the spatial radius changes in time*; it is no longer static.[8]

This reasoning made it clear that the requirement of staticity is not sacrosanct. Once it is dropped, entirely new classes of solutions become available. These are precisely the new solutions that were discovered by Friedmann and Lemaître. These universes are *truly dynamical*; in contrast to de Sitter spacetime, they do not admit a time-translation symmetry even in patches! The clarity of the Friedmann-Lemaître papers is striking. Indeed, presentations in most contemporary texts on cosmology is the same as Lemaître's. Friedmann's papers contain illuminating comments/remarks on the underlying geometry, while Lemaître also comments on the physical implications. In particular, he has a detailed discussion of what we now call the cosmological red-shift, rooted in the expansion of the universe, including numerical values using the then best available data. This paper contains (in Equations (23) and (24)) what was called the Hubble law until 2018, when the International Astronomical Union voted to change the name to the Hubble-Lemaître law. From a contemporary perspective, then, the passage from the Einstein and de Sitter solutions to the truly dynamical universes with matter, discovered by Friedmann and Lemaître, was a huge transition –a truly major step that transformed our view of the large scale structure of the universe we inhabit. Indeed, the title of the Russian biography of Friedmann calls him *'The man who made the universe expand'*.

Surprisingly, this impressive leap was not immediately hailed as an important breakthrough! Einstein first thought that dynamical solutions were not possible and, in a short 1922 note in Zeitschrift für Physik, he argued that Friedmann had made a mistake. He retracted his objection in another, shorter note in 1923 after he received an explanatory letter from Friedmann through his St. Petersburg colleague, Yuri Krutkov. However, he continued to think of these solutions merely as mathematical possibilities, having little relevance to the physical world. Indeed in 1927, while taking a stroll in the park Léopold during a break in the Solvay conference, he complimented Lemaître on his mathematical competence but also said that from a physical point of view he found the idea of an expanding universe *"tout à fait abominable"* (completely abominable)! It was a true uphill battle for the expanding universe of Friedman and Lemaître to be accepted by the wider community. The 1929 review that Einstein wrote for Encyclopedia Britanica, for example, makes no

---

[8]In this 1925 'Note on de Sitter Universe' published in the Journal of Mathematics and Physics, he also obtained spatially flat solutions but, unfortunately, rejected them on philosophical grounds. Note also that Lemaître's 1927 paper appeared in French in a rather obscure Belgian journal. The version included in this collection is the English translation that appeared in Monthly Notices of the Royal Astronomical Society at Eddington's initiative. Unfortunately, this translation omits an important para on red-shifts.

reference to the possibility of an expanding universe. A push in favor of an expanding universe came from Eddington's finding in 1930 –soon after Lemaître reminded him of his 1927 paper– that Einstein's universe is unstable: under slightest perturbations, it begins to expand (or contract). As a result, Eddington started to favor the idea of an expanding universe. It was only in 1931 after meeting Edwin Hubble in Pasadena that Einstein fully conceded that his static universe was not viable. However, in the report to the Prussian Academy on his return from the trip, he gave instability as the primary reason for his change of heart, *"quite apart from Hubble's observational results"*. And Einstein begins this discussion by paying a tribute to Friedmann: *"The first who, uninfluenced by observations, tried this way was A. Friedmann, on whose mathematical results the following remarks will be based."* Eddington and de Sitter came to fully appreciate Lemaître's work and, starting from 1930, emphasized its importance in interpreting Hubble's redshift observations as the discovery of an expanding universe. Einstein also embraced the expanding universe in a paper he wrote with de Sitter in 1932, where they considered the spatially flat expanding universe with *zero* λ; zero because Einstein believed and emphasized that, once the requirement of staticity is abandoned, the presence of a cosmological constant served no purpose, and its introduction in 1917 was a mistake.

In spite of these endorsements, the notion of expanding universe did not blossom in the subsequent years as one might have expected. Friedmann passed away in 1925 –just three years after making his seminal discovery. Lemaître continued to develop the notion further and arrived at the conclusion that a universe that expands in the future must start with, what Fred Hoyle later called, a Big Bang. Lemaître pursued the theme of a finite beginning systematically, starting with an article in Nature entitled 'The Beginning of the World from the Point of View of Quantum Theory.' However, leading figures were put off by it. Eddington, for example, publicly called the idea of a finite beginning *"repugnant."* In the 1940s the British school put forward 'Steady State Cosmology' as an alternative to the Big Bang and the expanding universe, returning, once again, to the age-old view that on a large scale the universe is unchanging, without a beginning or an end. In the mainstream cosmology, acceptance of the notion of an expanding universe –with an early hot phase– came only after the cosmic microwave background was discovered, and understanding of nuclear abundances matured. However, for several subsequent decades, the community continued to follow Einstein and set the cosmological constant λ to zero. Lemaître, on the other hand, remained committed to a positive λ throughout his life and tried to convince Einstein that a cosmological constant was necessary to solve what was then called the *"time scale dilemma."* In 1934, he clearly understood that, in its essence, λ *"corresponds to a negative density of vacuum ρ_0 according to $\rho_0 = (\lambda c^2)/4G$"* and that *"in order that motion relative to vacuum may not be detected, we must associate a pressure $p = \rho c^2$ to the vacuum."* He had interesting correspondence with Einstein on the issue of λ. In a 1947 letter, Einstein admitted that his preference for a zero value was based on aesthetics and added *"I am unable to think that such an ugly thing should be realized in Nature."* Some 45 years later, the supernovae observations showed that Nature is not easily swayed by aesthetic considerations of humans – even of the deepest thinkers among them! Ultimately, Lemaître's view did triumph, but he had already passed away in 1966.

The three parts of this collection, spanning the first quarter of the 20th century, highlight the critical transitions that profoundly altered our view of spacetime. They led us to a conceptual framework that underlies all of fundamental physics –from quarks to the cosmos– even today, almost a century later!

Abhay Ashtekar
State College, PA, USA
25th October 2020

Additional References

Papers included in this collection led to three major paradigm shifts in our under-standing of the structure of spacetime. The first set culminated in a fusion of space and time into a 4-dimensional spacetime continuum; the second led us to abandon the notion that gravity as a force exerted by massive bodies and to encode it, instead, in the very geometry of spacetime; and the third forced us to renounce the age-old notion that the universe is serene and time-unchanging on large scales. There are several other papers that have had significant impact on the contemporary notion of spacetime as it pertains to the large scale structure of our universe. We include a partial list of these early papers that readers may also want to consult.

G. Lemaître, "Note on de Sitter's universe", Journal of Mathematics and Physics (1925) 4, pp.188-192; Phys. Rev. (1925) 25, Ser. II, 903

E. P. Hubble, "A relation between distance and radial velocity among extra galac-tic nebulae", Proc. Nat. Acad. Sc. (1929) 15, pp.168-173

E. P. Hubble and M.L. Humason, "The velocity-distance relation among extra galactic nebulae" Ap. J. (1931) 74, pp. 43-80

A. Einstein and W. de Sitter, "On the Relation Between the Expansion and the Mean Density of the Universe", Proc. Nat. Acad. Sci. (1932) 18, pp.213-214

G. Lemaître, "The beginning of the world from the point of view of quantum theory" Nature (1931) 127, 706, Reproduced as a Golden Oldie and Editorial Note by J.-P. Luminet, Gen. Rel. Grav. (2011) 43, 2911-2918 (2011); "Evolution of the expanding universe" Proc. Nat. Acad. Sc. (1934) 20, pp.12-17

H. P. Robertson, "Kinematics and world structure", Astrophysical Journal (1935) 82, pp. 284-301; "Kinematics and world structure II", Astrophysical Journal (1936) 83, pp. 187-201; "Kinematics and world structure III", Astrophysical Journal (1936) 83, pp. 257-271

A. G. Walker, "On Milne's theory of world-structure", Proceedings of the London Mathematical Society, Series 2 (1937) 42, pp. 90-127

CONTENTS

III Birth of a Dynamical Universe   179

A. Einstein

Cosmological Considerations on the General Theory of Relativity 181

W. de Sitter

On the Curvature of Space   191

Part I

EMERGENCE OF THE SPACETIME CONTINUUM

MICHELSON'S INTERFERENCE EXPERIMENT

H. A. LORENTZ

Translation of Versuch einer Theorie der elektrischen und optischen Erscheinungen in bewegten Körpern, Leiden, 1895, §§ 89-92. Original English publication in: H. A. Lorentz, A. Einstein, H. Minkowski and H. Weyl, *The Principle of Relativity: A Collection of Original Memoirs on the Special and General Theory of Relativity.* With Notes by A. Sommerfeld. Translated by W. Perrett and G. B. Jeffery (Methuen and Company, Ltd., 1923; reprinted by Dover Publications Inc., 1952).

1. As Maxwell first remarked and as follows from a very simple calculation, the time required by a ray of light to travel from a point A to a point B and back to A must vary when the two points together undergo a displacement without carrying the ether with them. The difference is, certainly, a magnitude of second order; but it is sufficiently great to be detected by a sensitive interference method.

The experiment was carried out by Michelson in 1881.[1] His apparatus, a kind of interferometer, had two horizontal arms, P and Q, of equal length and at right angles one to the other. Of the two mutually interfering rays of light the one passed along the arm P and back, the other along the arm Q and back. The whole instrument, including the source of light and the arrangement for taking observations, could be revolved about a vertical axis; and those two positions come especially under consideration in which the arm P or the arm Q lay as nearly as possible in the direction of the Earth's motion. On the basis of Fresnel's theory it was anticipated that when the apparatus was revolved from one of these *principal positions* into the other there would be a displacement of the interference fringes.

But of such a displacement – for the sake of brevity we will call it the Maxwell displacement – conditioned by the change in the times of propagation, no trace was discovered, and accordingly Michelson thought himself justified in concluding that while the Earth is moving, the ether does not remain at rest. The correctness of this inference was soon brought into question, for by an oversight Michelson had taken the change in the phase difference, which was to be expected in accordance with the theory, at twice its proper value. If we make the necessary correction, we arrive at displacements no greater than might be masked by errors of observation.

Subsequently Michelson[2] took up the investigation anew in collaboration with Morley, enhancing the delicacy of the experiment by causing each pencil to be re-

---

[1]Michelson, American Journal of Science, 22, 1881, p. 120.

[2]Michelson and Morley, American Journal of Science, 34, 1887, p. 333; Phil. Mag., 24, 1887, p. 449.

flected to and fro between a number of mirrors, thereby obtaining the same advantage as if the arms of the earlier apparatus had been considerably lengthened. The mirrors were mounted on a massive stone disc, floating on mercury, and therefore easily revolved. Each pencil now had to travel a total distance of 22 meters, and on Fresnel's theory the displacement to be expected in passing from the one principal position to the other would be 0.4 of the distance between the interference fringes. Nevertheless the rotation produced displacements not exceeding 0.02 of this distance, and these might well be ascribed to errors of observation.

Now, does this result entitle us to assume that the ether takes part in the motion of the Earth, and therefore that the theory of aberration given by Stokes is the correct one? The difficulties which this theory encounters in explaining aberration seem too great for me to share this opinion, and I would rather try to remove the contradiction between Fresnel's theory and Michelson's result. An hypothesis which I brought forward some time ago,[3] and which, as I subsequently learned, has also occurred to Fitzgerald,[4] enables us to do this. The next paragraph will set out this hypothesis.

2. To simplify matters we will assume that we are working with apparatus as employed in the first experiments, and that in the one principal position the arm P lies exactly in the direction of the motion of the Earth. Let v be the velocity of this motion, L the length of either arm, and hence $2L$ the path traversed by the rays of light. According to the theory[5] the turning of the apparatus through 90° causes the time in which the one pencil travels along P and back to be longer than the time which the other pencil takes to complete its journey by

$$\frac{Lv^2}{c^2}.$$

There would be this same difference if the translation had no influence and the arm P were longer than the arm Q by $\frac{1}{2}Lv^2/c^2$. Similarly with the second principal position.

Thus we see that the phase differences expected by the theory might also arise if, when the apparatus is revolved, first the one arm and then the other arm were the longer. It follows that the phase differences can be compensated by contrary changes of the dimensions.

If we assume the arm which lies in the direction of the Earth's motion to be shorter than the other by $\frac{1}{2}Lv^2/c^2$, and, at the same time, that the translation has the influence which Fresnel's theory allows it, then the result of the Michelson experiment is explained completely.

Thus one would have to imagine that the motion of a solid body (such as a brass rod or the stone disc employed in the later experiments) through the resting ether exerts upon the dimensions of that body an influence which varies according to the orientation of the body with respect to the direction of motion. If, for example, the dimensions parallel to this direction were changed in the proportion of 1 to $1 + \delta$, and

---

[3]Lorentz, Zittingsverslagen der Akad. v. Wet. te Amsterdam, 1892-93, S. 74.

[4]As Fitzgerald kindly tells me, he has for a long time dealt with his hypothesis in his lectures. The only published reference which I can find to the hypothesis is by Lodge, "Aberration Problems," Phil. Trans. R. S., 184 A, 1893, S. 727.

[5]Cf. Lorentz, Arch. Néerl., 2, 1887, pp. 168-176.

those perpendicular in the proportion of 1 to $1 + \epsilon$, then we should have the equation

$$\epsilon - \delta = \frac{1}{2}\frac{v^2}{c^2},\tag{1}$$

in which the value of one of the quantities δ and ϵ would remain undetermined. It might be that $\epsilon = 0, \delta = -\frac{1}{2}v^2/c^2$ but also $\epsilon = \frac{1}{2}v^2/c^2, \delta = 0$, or $\epsilon = \frac{1}{4}v^2/c^2$, and $\delta = -\frac{1}{4}v^2/c^2$.

3. Surprising as this hypothesis may appear at first sight, yet we shall have to admit that it is by no means far-fetched, as soon as we assume that molecular forces are also transmitted through the ether, like the electric and magnetic forces of which we are able at the present time to make this assertion definitely. If they are so transmitted, the translation will very probably affect the action between two molecules or atoms in a manner resembling the attraction or repulsion between charged particles. Now, since the form and dimensions of a solid body are ultimately conditioned by the intensity of molecular actions, there cannot fail to be a change of dimensions as well.

From the theoretical side, therefore, there would be no objection to the hypothesis. As regards its experimental proof, we must first of all note that the lengthenings and shortenings in question are extraordinarily small. We have $v^2/c^2 = 10^{-8}$, and thus, if $\epsilon = 0$, the shortening of the one diameter of the Earth would amount to about 6.5 cm. The length of a meter rod would change, when moved from one principal position into the other, by about 1/200 microns. One could hardly hope for success in trying to perceive such small quantities except by means of an interference method. We should have to operate with two perpendicular rods, and with two mutually interfering pencils of light, allowing the one to travel to and fro along the first rod, and the other along the second rod. But in this way we should come back once more to the Michelson experiment, and revolving the apparatus we should perceive no displacement of the fringes. Reversing a previous remark, we might now say that the displacement produced by the alterations of length is compensated by the Maxwell displacement.

4. It is worth noticing that we are led to just the same changes of dimensions as have been presumed above if we, *firstly*, without taking molecular movement into consideration, assume that in a solid body left to itself the forces, attractions or repulsions, acting upon any molecule maintain one another in equilibrium, and, *secondly* – though to be sure, there is no reason for doing so – if we apply to these molecular forces the law which in another place[6] we deduced for electrostatic actions. For if we now understand by S_1 and S_2 not, as formerly, two systems of charged particles, but two systems of molecules – the second at rest and the first moving with a velocity v in the direction of the axis of x – between the dimensions of which the relationship subsists as previously stated; and if we assume that in both systems the x components of the forces are the same, while the y and z components differ from one another by the factor $\sqrt{1 - v^2/c^2}$, then it is clear that the forces in S_1 will be in equilibrium whenever they are so in S_2. If therefore S_2 is the state of equilibrium of a solid body

---

[6]Viz., § 23 of the book, "Versuch einer Theorie der elektrischen und optischen Erscheinungen in bewegten Korpern."

at rest, then the molecules in S_1 have precisely those positions in which they can persist under the influence of translation. The displacement would naturally bring about this disposition of the molecules of its own accord, and thus effect a shortening in the direction of motion in the proportion of 1 to $\sqrt{1 - v^2/c^2}$, in accordance with the formulae given in the above-mentioned paragraph. This leads to the values

$$\delta = -\frac{1}{2}\frac{v^2}{c^2}, \epsilon = 0$$

in agreement with (1).

In reality the molecules of a body are not at rest, but in every "state of equilibrium" there is a stationary movement. What influence this circumstance may have in the phenomenon which we have been considering is a question which we do not here touch upon; in any case the experiments of Michelson and Morley, in consequence of unavoidable errors of observation, afford considerable latitude for the values of δ and ϵ .

ELECTROMAGNETIC PHENOMENA IN A SYSTEM MOVING WITH ANY VELOCITY SMALLER THAN THAT OF LIGHT

H. A. LORENTZ

New publication of the English version in Proceedings of the Academy of Sciences of Amsterdam, 6, 1904. Original English publication in: H. A. Lorentz, A. Einstein, H. Minkowski and H. Weyl, *The Principle of Relativity: A Collection of Original Memoirs on the Special and General Theory of Relativity.* With Notes by A. Sommerfeld. Translated by W. Perrett and G. B. Jeffery (Methuen and Company, Ltd., 1923; reprinted by Dover Publications Inc., 1952).

§ 1. The problem of determining the influence exerted on electric and optical phenomena by a translation, such as all systems have in virtue of the Earth's annual motion, admits of a comparatively simple solution, so long as only those terms need be taken into account, which are proportional to the first power of the ratio between the velocity of translation v and the velocity of light c. Cases in which quantities of the second order, i.e. of the order v^2/c^2 may be perceptible, present more difficulties. The first example of this kind is Michelson's well-known interference experiment, the negative result of which has led Fitzgerald and myself to the conclusion that the dimensions of solid bodies are slightly altered by their motion through the ether.

Some new experiments, in which a second order effect was sought for, have recently been published. Rayleigh[1] and Brace[2] have examined the question whether the Earth's motion may cause a body to become doubly refracting. At first sight this might be expected, if the just mentioned change of dimensions is admitted. Both physicists, however, have obtained a negative result.

In the second place Trouton and Noble[3] have endeavoured to detect a turning couple acting on a charged condenser, the plates of which make a certain angle with the direction of translation. The theory of electrons, unless it be modified by some new hypothesis, would undoubtedly require the existence of such a couple. In order to see this, it will suffice to consider a condenser with ether as dielectric. It may be shown that in every electrostatic system, moving with a velocity v,[4] there is a

---

[1] Rayleigh, Phil. Mag. (6), 4,1902, p. 678.

[2] Brace, Phil. Mag. (6), 7, 1904, p. 317.

[3] Trouton and Noble, Phil. Trans. Roy. Soc. Lond., A 202, 1903, p. 165.

[4] A vector will be denoted by a Clarendon letter, its magnitude by the corresponding Latin letter.

certain amount of "electromagnetic momentum." If we represent this, in direction and magnitude, by a vector G, the couple in question will be determined by the vector product[5]

$$[\mathbf{G} \cdot \mathbf{v}] \tag{1}$$

Now, if the axis of z is chosen perpendicular to the condenser plates, the velocity v having any direction we like; and if U is the energy of the condenser, calculated in the ordinary way, the components of G are given[6] by the following formulae, which are exact up to the first order,

$$G_x = \frac{2U}{c^2} v_x, \qquad G_y = \frac{2U}{c^2} v_y, \qquad G_z = 0.$$

Substituting these values in (1), we get for the components of the couple, up to terms of the second order,

$$\frac{2U}{c^2} v_y v_z, \qquad -\frac{2U}{c^2} v_x v_z, \qquad 0.$$

These expressions show that the axis of the couple lies in the plane of the plates, perpendicular to the translation. If α is the angle between the velocity and the normal to the plates, the moment of the couple will be $U(v/c)^2 \sin 2\alpha$; it tends to turn the condenser into such a position that the plates are parallel to the Earth's motion.

In the apparatus of Trouton and Noble the condenser was fixed to the beam of a torsion-balance, sufficiently delicate to be deflected by a couple of the above order of magnitude. No effect could however be observed.

§ 2. The experiments of which I have spoken are not the only reason for which a new examination of the problems connected with the motion of the Earth is desirable. Poincaré[7] has objected to the existing theory of electric and optical phenomena in moving bodies that, in order to explain Michelson's negative result, the introduction of a new hypothesis has been required, and that the same necessity may occur each time new facts will be brought to light. Surely this course of inventing special hypotheses for each new experimental result is somewhat artificial. It would be more satisfactory if it were possible to show by means of certain fundamental assumptions and without neglecting terms of one order of magnitude or another, that many electromagnetic actions are entirely independent of the motion of the system. Some years ago, I already sought to frame a theory of this kind.[8] I believe it is now possible to treat the subject with a better result. The only restriction as regards the velocity will be that it be less than that of light.

§ 3. I shall start from the fundamental equations of the theory of electrons.[9] Let $\mathbf{D}$ be the dielectric displacement in the ether, $\mathbf{H}$ the magnetic force, ρ the volume-density of the charge of an electron, $\mathbf{v}$ the velocity of a point of such a particle, and

---

[5] See my article : "Weiterbildung der Maxwell'schen Theorie. Electronentheorie," Mathem. Encyclopadie, V, 14, § 21, a. (This article will be quoted as " M.E.")

[6] "M.E.," § 56, c.

[7] Poincaré, Rapports du Congrès de physique de 1900, Paris, 1, pp. 22, 23.

[8] Lorentz, Zittingsverslag Akad. v. Wet., 7, 1899, p. 507; Amsterdam Proc., 1898-99, p. 427.

[9] "M.E.," § 2.

F the ponderomotive force, i.e. the force, reckoned per unit charge, which is exerted by the ether on a volume-element of an electron. Then, if we use a fixed system of coordinates,

$$\operatorname{div} \mathbf{D} = \rho, \ \operatorname{div} \mathbf{H} = 0,$$

$$\operatorname{curl} \mathbf{H} = \frac{1}{c}\left(\frac{\partial \mathbf{D}}{\partial t} + \rho \mathbf{v}\right),$$

$$\operatorname{curl} \mathbf{D} = -\frac{1}{c}\frac{\partial \mathbf{H}}{\partial t},$$

$$\mathbf{F} = \mathbf{D} + \frac{1}{c}[\mathbf{v} \cdot \mathbf{H}].$$

(2)

I shall now suppose that the system as a whole moves in the direction of x with a constant velocity v, and I shall denote by **u** any velocity which a point of an electron may have in addition to this, so that

$$v_x = v + u_x, \qquad v_y = u_y, \qquad v_z = u_z.$$

If the equations (2) are at the same time referred to axes moving with the system, they become

$$\operatorname{div} \mathbf{D} = \rho, \qquad \operatorname{div} \mathbf{H} = 0,$$

$$\frac{\partial H_z}{\partial y} - \frac{\partial H_y}{\partial z} = \frac{1}{c}\left(\frac{\partial}{\partial t} - v\frac{\partial}{\partial x}\right)D_x + \frac{1}{c}\rho(v + u_x),$$

$$\frac{\partial H_x}{\partial z} - \frac{\partial H_z}{\partial x} = \frac{1}{c}\left(\frac{\partial}{\partial t} - v\frac{\partial}{\partial x}\right)D_y + \frac{1}{c}\rho u_y,$$

$$\frac{\partial H_y}{\partial x} - \frac{\partial H_x}{\partial y} = \frac{1}{c}\left(\frac{\partial}{\partial t} - v\frac{\partial}{\partial x}\right)D_z + \frac{1}{c}\rho u_z,$$

$$\frac{\partial D_z}{\partial y} - \frac{\partial D_y}{\partial z} = -\frac{1}{c}\left(\frac{\partial}{\partial t} - v\frac{\partial}{\partial x}\right)H_x,$$

$$\frac{\partial D_x}{\partial z} - \frac{\partial D_z}{\partial x} = -\frac{1}{c}\left(\frac{\partial}{\partial t} - v\frac{\partial}{\partial x}\right)H_y,$$

$$\frac{\partial D_y}{\partial x} - \frac{\partial D_x}{\partial y} = -\frac{1}{c}\left(\frac{\partial}{\partial t} - v\frac{\partial}{\partial x}\right)H_z,$$

$$F_x = D_x + \frac{1}{c}(u_y H_z - u_z H_y),$$

$$F_y = D_y - \frac{1}{c}v H_z + \frac{1}{c}(u_z H_x - u_x H_z),$$

$$F_z = D_z + \frac{1}{c}v H_y + \frac{1}{c}(u_x H_y - u_y H_x).$$

§ 4. We shall further transform these formulae by a change of variables. Putting

$$\frac{c^2}{c^2 - v^2} = \beta^2,$$

(3)

and understanding by l another numerical quantity, to be determined further on, I take as new independent variables

$$x' = \beta l x, \qquad y' = l y, \qquad z' = l z, \tag{4}$$

$$t' = \frac{l}{\beta} t - \beta l \frac{v}{c^2} x, \tag{5}$$

and I define two new vectors $\mathbf{D}'$ and $\mathbf{H}'$ by the formulae

$$D'_x = \frac{1}{l^2} D_x, \quad D'_y = \frac{\beta}{l^2}\left(D_y - \frac{v}{c} H_z\right), \quad D'_z = \frac{\beta}{l^2}\left(D_z + \frac{v}{c} H_y\right),$$
$$H'_x = \frac{1}{l^2} H_x, \quad H'_y = \frac{\beta}{l^2}\left(H_y + \frac{v}{c} D_z\right), \quad H'_z = \frac{\beta}{l^2}\left(H_z - \frac{v}{c} D_y\right),$$

for which, on account of (3), we may also write

$$
\left.
\begin{aligned}
D_x &= l^2 D'_x, \quad D_y = \beta l^2\left(D'_y + \frac{v}{c} H'_z\right), \quad D_z = \beta l^2\left(D'_z - \frac{v}{c} H'_y\right) \\
H_x &= l^2 H'_x, \quad H_y = \beta l^2\left(H'_y - \frac{v}{c} D'_z\right), \quad H_z = \beta l^2\left(H'_z + \frac{v}{c} D'_y\right).
\end{aligned}
\right\} \tag{6}
$$

As to the coefficient l, it is to be considered as a function of v, whose value is 1 for $v = 0$, and which, for small values of v, differs from unity no more than by a quantity of the second order.

The variable t' may be called the "local time;" indeed, for $\beta = 1, l = 1$, it becomes identical with what I formerly denoted by this name.

If, finally, we put

$$\frac{1}{\beta l^3} \rho = \rho' \tag{7}$$

$$\beta^2 u_x = u'_x, \qquad \beta u_y = u'_y, \qquad \beta u_z = u'_z, \tag{8}$$

these latter quantities being considered as the components of a new vector $\mathbf{u}'$, the equations take the following form

$$\operatorname{div}' \mathbf{D}' = \left(1 - \frac{v u'_x}{c^2}\right)\rho', \quad \operatorname{div}' \mathbf{H}' = 0,$$

$$\operatorname{curl}' \mathbf{H}' = \frac{1}{c}\left(\frac{\partial \mathbf{D}'}{\partial t'} + \rho' \mathbf{u}'\right), \tag{9}$$

$$\operatorname{curl}' \mathbf{D}' = -\frac{1}{c}\frac{\partial \mathbf{H}'}{\partial t'},$$

$$F_x = l^2 \left\{ D'_x + \frac{1}{c}(u'_y H'_z - u'_z H'_y) + \frac{v}{c^2}(u'_y D'_y + u'_z D'_z) \right\},$$

$$F_y = \frac{l^2}{\beta} \left\{ D'_y + \frac{1}{c}(u'_z H'_x - u'_x H'_z) - \frac{v}{c^2} u'_x D'_y \right\}, \tag{10}$$

$$F_z = \frac{l^2}{\beta} \left\{ D'_z + \frac{1}{c}(u'_x H'_y - u'_y H'_x) - \frac{v}{c^2} u'_x D'_z \right\}.$$

The meaning of the symbols div′ and curl′ in (9) is similar to that of div and curl in (2); only, the differentiations with respect to x, y, z are to be replaced by the corresponding ones with respect to x', y', z'.

§ 5. The equations (9) lead to the conclusion that the vectors D' and H' may be represented by means of a scalar potential ϕ and a vector potential A'. These potentials satisfy the equations[10]

$$\nabla'^2 \phi' - \frac{1}{c^2}\frac{\partial^2 \phi'}{\partial t'^2} = -\rho' \tag{11}$$

$$\nabla'^2 A' - \frac{1}{c^2}\frac{\partial^2 A'}{\partial t'^2} = -\frac{1}{c}\rho' u', \tag{12}$$

and in terms of them $\mathbf{D'}$ and $\mathbf{H'}$ are given by

$$\mathbf{D'} = -\frac{1}{c}\frac{\partial \mathbf{A'}}{\partial t'} - \operatorname{grad}' \phi' + \frac{v}{c}\operatorname{grad}' A'_x \tag{13}$$

$$\mathbf{H'} = \operatorname{curl}' \mathbf{A'}. \tag{14}$$

The symbol ∇'^2 is an abbreviation for

$$\frac{\partial^2}{\partial x'^2} + \frac{\partial^2}{\partial y'^2} + \frac{\partial^2}{\partial z'^2}$$

and $\operatorname{grad}' \phi'$ denotes a vector whose components are

$$\frac{\partial \phi'}{\partial x'}, \qquad \frac{\partial \phi'}{\partial y'}, \qquad \frac{\partial \phi'}{\partial z'},$$

The expression $\operatorname{grad}' A'_x$ has a similar meaning.

In order to obtain the solution of (11) and (12) in a simple form, we may take x', y', z' as the coordinates of a point P' in a space S', and ascribe to this point, for each value of t', the values of ρ', u', ϕ', A', belonging to the corresponding point $P(x, y, z)$ of the electromagnetic system. For a definite value t' of the fourth independent variable, the potentials ϕ' and A' at the point P of the system or at the corresponding point P' of the space S', are given by[11]

$$\phi' = \frac{1}{4\pi}\int \frac{[\rho']}{r'}dS' \tag{15}$$

$$\mathbf{A'} = \frac{1}{4\pi c}\int \frac{[\rho' \mathbf{u'}]}{r'}dS'. \tag{16}$$

Here dS' is an element of the space S', r' its distance from P', and the brackets serve to denote the quantity ρ' and the vector $\rho' \mathbf{u'}$ such as they are in the element dS' , for the value $t' - r'/c$ of the fourth independent variable.

---

[10]"M.E.," §§ 4 and 10.

[11] *Ibid.*, §§ 5 and 10.

Instead of (15) and (16) we may also write, taking into account (4) and (7),

$$\phi' = \frac{1}{4\pi} \int \frac{[\rho]}{r} dS \tag{17}$$

$$\mathbf{A}' = \frac{1}{4\pi c} \int \frac{[\rho \mathbf{u}]}{r} dS, \tag{18}$$

the integrations now extending over the electromagnetic system itself. It should be kept in mind that in these formulae r' does not denote the distance between the element dS and the point (z, y, z) for which the calculation is to be performed. If the element lies at the point (x_1, y_1, z_1) we must take

$$r' = l\sqrt{\beta^2(x - x_2)^2 + (y - y_1)^2 + (z - z_1)^2}.$$

It is also to be remembered that, if we wish to determine ϕ' and $\mathbf{A}'$ for the instant at which the local time in P is t' we must take ρ and $\rho\mathbf{u}'$, such as they are in the element dS at the instant at which the local time of that element is $t' - r'/c$.

§ **6.** It will suffice for our purpose to consider two special cases The first is that of an electrostatic system i.e. a system having no other motion but the translation with the velocity v. In this case $u' = 0$, and therefore, by (12), $\mathbf{A}' = 0$. Also, ϕ' is independent of t', so that the equations (11), (13), and (14) reduce to

$$\nabla'^2 \phi' = -\rho',$$
$$\mathbf{D}' = -\operatorname{grad}', \phi', \tag{19}$$
$$\mathbf{H}' = 0.$$

After having determined the vector $\mathbf{D}'$ by means of these equations we know also the ponderomotive force acting on electrons that belong to the system. For these the formulae (10) become, since $\mathbf{u}' = 0$

$$F_x = l^2 D_x', \qquad F = \frac{l^2}{\beta} D_y', \qquad F_z = \frac{l^2}{\beta} D_z'. \tag{20}$$

The result may be put in a simple form if we compare the moving system Σ, with which we are concerned, to another electrostatic system Σ' which remains at rest, and into which Σ is changed if the dimensions parallel to the axis of x are multiplied by βl, and the dimensions which have the direction of y or that of z, by $l - a$ deformation for which $(\beta l, l, l)$ is an appropriate symbol. In this new system, which we may suppose to be placed in the above-mentioned space S', we shall give to the density the value ρ', determined by (7), so that the charges of corresponding elements of volume and of corresponding electrons are the same in Σ and Σ'. Then we shall obtain the forces acting on the electrons of the moving system Σ, if we first determine the corresponding forces in Σ', and next multiply their components in the direction of the axis of x by l^2, and their components perpendicular to that axis by l^2/β. This is conveniently expressed by the formula

$$\mathbf{F}(\Sigma) = \left(l^2, \frac{l^2}{\beta}, \frac{l^2}{\beta}\right) \mathbf{F}(\Sigma'). \tag{21}$$

It is further to be remarked that, after having found $\mathbf{D}'$ by (19), we can easily calculate the electromagnetic momentum in the moving system, or rather its component in the direction of the motion. Indeed, the formula

$$\mathbf{G} = \frac{1}{c} \int [\mathbf{D.H}]\, dS$$

shows that

$$G_x = \frac{1}{c} \int (D_y H_z - D_z H_y) dS.$$

Therefore, by (6), since $\mathbf{H}' = 0$

$$G_x = \frac{\beta^2 l^4 v}{c^2} \int (D_y^{'2} + D_z^{'2}) dS = \frac{\beta l v}{c^2} \int (D_y^{'2} + D_z^{'2}) dS'. \tag{22}$$

§ 7. Our second special case is that of a particle having an electric moment, i.e. a small space S, with a total charge

$$\int \rho\, dS = 0$$

but with such a distribution of density that the integrals

$$\int \rho x\, dS, \int \rho y\, dS, \int \rho z\, dS$$

have values differing from 0. Let ξ, μ, ζ, be the coordinates, taken relatively to a fixed point A of the particle, which may be called its centre, and let the electric moment be defined as a vector P whose components are

$$P_x = \int \rho\xi\, dS, \qquad P_y = \int \rho\eta\, dS, \qquad P_z = \int \rho\zeta\, dS. \tag{23}$$

Then

$$\frac{dP_x}{dt} = \int \rho u_x\, dS, \qquad \frac{dP_y}{dt} = \int \rho u_y\, dS, \qquad \frac{dP_z}{dt} = \int \rho u_z\, dS. \tag{24}$$

Of course, if ξ, η, ζ are treated as infinitely small, u_x, u_y, u_z must be so likewise. We shall neglect squares and products of these six quantities.

We shall now apply the equation (17) to the determination of the scalar potential ϕ' for an exterior point $P(x, y, z)$ at a finite distance from the polarized particle, and for the instant at which the local time of this point has some definite value t'. In doing so, we shall give the symbol $[\rho]$, which, in (17), relates to the instant at which the local time in dS is $t' - r'/c$, a slightly different meaning. Distinguishing by r_0' the value of r' for the centre A, we shall understand by $[\rho]$ the value of the density existing in the element dS at the point ξ, η, ζ at the instant t_0 at which the local time of A is $t' - r_0/c$.

It may be seen from (5) that this instant precedes that for which we have to take the numerator in (17) by

$$\beta^2 \frac{v\xi}{c^2} + \frac{\beta(r_0' - r')}{lc} = \beta^2 \frac{v\xi}{c^2} + \frac{\beta}{lc}\left(\xi \frac{\partial r'}{\partial x} + \eta \frac{\partial r'}{\partial y} + \zeta \frac{\partial r'}{\partial z}\right)$$

units of time. In this last expression we may put for the differential coefficients their values at the point A.

In (17) we have now to replace $[\rho]$ by

$$[\rho] + \beta^2 \frac{v\xi}{c^2} \left[\frac{\partial\rho}{\partial t}\right] + \frac{\beta}{lc}\left(\xi\frac{\partial r'}{\partial x} + \eta\frac{\partial r'}{\partial y} + \zeta\frac{\partial r'}{\partial z}\right)\left[\frac{\partial\rho}{\partial t}\right], \tag{25}$$

where $\left[\frac{\partial\rho}{\partial t}\right]$ relates again to the time t_0. Now, the value of t' for which the calculations are to be performed having been chosen, this time t_0 will be a function of the coordinates x, y, z of the exterior point P. The value of $[\rho]$ will therefore depend on these coordinates in such a way that

$$\frac{\partial[\rho]}{\partial x} = -\frac{\beta}{lc}\frac{\partial r'}{\partial x}\left[\frac{\partial\rho}{\partial t}\right], \text{ etc.}$$

by which (25) becomes

$$[\rho] + \beta^2 \frac{v\xi}{c^2}\left[\frac{\partial\rho}{\partial t}\right] - \left(\xi\frac{\partial[\rho]}{\partial x} + \eta\frac{\partial[\rho]}{\partial y} + \zeta\frac{\partial[\rho]}{\partial z}\right).$$

Again, if henceforth we understand by r' what has above been called r'_0, the factor $\frac{1}{r'}$ must be replaced by

$$\frac{1}{r'} - \xi\frac{\partial}{\partial x}\left(\frac{1}{r'}\right) - \eta\frac{\partial}{\partial y}\left(\frac{1}{r'}\right) - \zeta\frac{\partial}{\partial z}\left(\frac{1}{r'}\right),$$

so that after all, in the integral (17), the element dS is multiplied by

$$\frac{[\rho]}{r'} + \frac{\beta^2 v\xi}{c^2 r'}\left[\frac{\partial\rho}{\partial t}\right] - \frac{\partial}{\partial x}\frac{\xi[\rho]}{r'} - \frac{\partial}{\partial y}\frac{\eta[\rho]}{r'} - \frac{\partial}{\partial z}\frac{\zeta[\rho]}{r'}.$$

This is simpler than the primitive form, because neither r' nor the time for which the quantities enclosed in brackets are to be taken, depend on x, y, z. Using (23) and remembering that $\int \rho\, dS = 0$, we get

$$\phi' = \frac{\beta^2 v}{4pc^2 r'}\left[\frac{\partial P_x}{\partial t}\right] - \frac{1}{4p}\left\{\frac{\partial}{\partial x}\frac{[P_x]}{r'} + \frac{\partial}{\partial y}\frac{[P_y]}{r'} + \frac{\partial}{\partial z}\frac{[P_z]}{r'}\right\},$$

a formula in which all the enclosed quantities are to be taken for the instant at which the local time of the centre of the particle is $t' - r'/c$.

We shall conclude these calculations by introducing a new vector $\mathbf{P}'$, whose components are

$$P'_x = \beta l P_x, \qquad P'_y = l P_y, \qquad P'_z = l P_z, \tag{26}$$

passing at the same time to x', y', z', t' as independent variables. The final result is

$$\phi' = \frac{v}{4pc^2 r'}\frac{\partial[P'_x]}{\partial t'} - \frac{1}{4p}\left\{\frac{\partial}{\partial x'}\frac{[P'_x]}{r'} + \frac{\partial}{\partial y'}\frac{[P'_y]}{r'} = \frac{\partial}{\partial z'}\frac{[P'_z]}{r'}\right\}.$$

As to the formula (18) for the vector potential, its transformation is less complicated, because it contains the indefinitely small vector $\mathbf{u}'$. Having regard to (8), (24), (26), and (5), I find

$$\mathbf{A}' = \frac{1}{4\pi cr'} \frac{\partial [\mathbf{P}']}{\partial t'}.$$

The field produced by the polarized particle is now wholly determined. The formula (13) leads to

$$\mathbf{D}' = -\frac{1}{4\pi c^2} \frac{\partial^2}{\partial t'^2} \frac{[\mathbf{P}']}{r'} + \frac{1}{4\pi} \mathrm{grad}' \left\{ \frac{\partial}{\partial x'} \frac{[P'_x]}{r'} + \frac{\partial}{\partial y'} \frac{[P'_y]}{r'} + \frac{\partial}{\partial z'} \frac{[P'_z]}{r'} \right\} \tag{27}$$

and the vector H' is given by (14). We may further use the equations (20), instead of the original formula (10), if we wish to consider the forces exerted by the polarized particle on a similar one placed at some distance. Indeed, in the second particle, as well as in the first, the velocities $\mathbf{u}$ may he held to be infinitely small.

It is to be remarked that the formulae for a system without translation are implied in what precedes. For such a system the quantities with accents become identical to the corresponding ones without accents; also $\beta = 1$ an $l = 1$. The components of (27) are at the same time those of the electric force which is exerted by one polarized particle

§ 8. Thus far we have used only the fundamental equations without any new assumptions. I shall now suppose *that the electrons, which I take to be spheres of radius R in the state of rest, have their dimensions changed by the effect of a translation, the dimensions in the direction of motion becoming βl times and those in perpendicular directions l times smaller.*

In this deformation, which may be represented by $\left(\frac{1}{\beta l}, \frac{1}{l}, \frac{1}{l}\right)$ each element of volume is understood to preserve its charge.

Our assumption amounts to saying that in an electrostatic system Σ, moving with a velocity v, all electrons are flattened ellipsoids with their smaller axes in the direction of motion. If now, in order to apply the theorem of § 6, we subject the system to the deformation $(\beta l, l, l)$, we shall have again spherical electrons of radius R. Hence, if we alter the relative position of the centres of the electrons in Σ by applying the deformation $(\beta l, l, l)$, and if, in the points thus obtained, we place the centres of electrons that remain at rest, we shall get a system, identical to the imaginary system Σ', of which we have spoken in § 6. The forces in this system and those in Σ will bear to each other the relation expressed by (21).

In the second place I shall suppose *that the forces between uncharged particles, as well as those between such particles and electrons, are influenced by a translation in quite the same way as the electric forces in an electrostatic system.* In other terms, whatever be the nature of the particles composing a ponderable body, so long as they do not move relatively to each other, we shall have between the forces acting in a system (Σ') without, and the same system (Σ) with a translation, the relation specified in (21), if, as regards the relative position of the particles, Σ' is got from Σ by the deformation $(\beta l, l, l)$ or Σ from Σ' by the deformation $\left(\frac{1}{\beta l}, \frac{1}{l}, \frac{1}{l}\right)$.

We see by this that, as soon as the resulting force is zero for a particle in Σ', the same must be true for the corresponding particle in Σ. Consequently, if, neglecting the effects of molecular motion, we suppose each particle of a solid body to be in equilibrium under the action of the attractions and repulsions exerted by its neighbours, and if we take for granted that there is but one configuration of equilibrium, we may draw the conclusion that the system Σ', if the velocity v is imparted to it, will of *itself* change into the system Σ. In other terms, the translation will *produce* the deformation $\left(\frac{1}{\beta l'}, \frac{1}{l'}, \frac{1}{l}\right)$.

The case of molecular motion will be considered in § 12.

It will easily be seen that the hypothesis which was formerly advanced in connection with Michelson's experiment, is implied in what has now been said. However, the present hypothesis is more general, because the only limitation imposed on the motion is that its velocity be less than that of light.

§ 9. We are now in a position to calculate the electromagnetic momentum of a single electron. For simplicity's sake I shall suppose the charge e to be uniformly distributed over the surface, so long as the electron remains at rest. Then a distribution of the same kind will exist in the system Σ with which we are concerned in the last integral of (22). Hence

$$\int (D'^2_y + D'^2_z)dS' = \frac{2}{3}\int D'^2 dS' = \frac{e^2}{6\pi}\int_R^\infty \frac{dr}{r^2} = \frac{e^2}{6\pi R},$$

and

$$G_x = \frac{e^2}{6\pi c^2 R}\beta l v.$$

It must be observed that the product βl is a function of v and that, for reasons of symmetry, the vector $\mathbf{G}$ has the direction of the translation. In general, representing by $\mathbf{v}$ the velocity of this motion, we have the vector equation

$$\mathbf{G} = \frac{e^2}{6\pi c^2 R}\beta l\mathbf{v}. \tag{28}$$

Now, every change in the motion of a system will entail a corresponding change in the electromagnetic momentum and will therefore require a certain force, which is given in direction and magnitude by

$$\mathbf{F} = \frac{d\mathbf{G}}{dt}. \tag{29}$$

Strictly speaking, the formula (28) may only be applied in the case of a uniform rectilinear translation. On account of this circumstance – though (29) is always true – the theory of rapidly varying motions of an electron becomes very complicated, the more so, because the hypothesis of § 8 would imply that the direction and amount of the deformation are continually changing. It is, indeed, hardly probable that the form of the electron will be determined solely by the velocity existing at the moment considered.

Nevertheless, provided the changes in the state of motion be sufficiently slow, we shall get a satisfactory approximation by using (28) at every instant. The application of (29) to such a *quasi-stationary* translation, as it has been called by Abraham,[12] is

---

[12] Abraham, Wied. Ann., 10, 1903, p. 105.

a very simple matter. Let, at a certain instant, $\mathbf{a_1}$ be the acceleration in the direction of the path, and $\mathbf{a_2}$ the acceleration perpendicular to it. Then the force $\mathbf{F}$ will consist of two components, having the directions of these accelerations and which are given by

$$\mathbf{F_1} = m_1\mathbf{a_1} \quad \text{and} \quad \mathbf{F_2} = m_2\mathbf{a_2},$$

if

$$m_1 = \frac{e^2}{6\pi c^2 R}\frac{d(\beta l v)}{dv} \quad \text{and} \quad m_2 = \frac{e^2}{6\pi c^2 R}\beta l,, \tag{30}$$

Hence, in phenomena in which there is an acceleration in the direction of motion, the electron behaves as if it had a mass m_1; in those in which the acceleration is normal to the path, as if the mass were m_2. These quantities m_1 and m_2 may therefore properly be called the "longitudinal" and "transverse " electromagnetic masses of the electron. I shall suppose *that there is no other, no "true" or "material" mass.*

Since β and l differ from unity by quantities of the order v^2/c^2 , we find for very small velocities

$$m_1 = m_2 = \frac{e^2}{6\pi c^2 R}.$$

This is the mass with which we are concerned, if there are small vibratory motions of the electrons in a system without translation. If, on the contrary, motions of this kind are going on in a body moving with the velocity v in the direction of the axis of x, we shall have to reckon with the mass m_1, as given by (30), if we consider the vibrations parallel to that axis, and with the mass m_2, if we treat of those that are parallel to OY or OZ. Therefore, in short terms, referring by the index Σ to a moving system and by Σ' to one that remains at rest,

$$m(\Sigma) = \left(\frac{d(\beta l v)}{dv}, \beta l, \beta l\right)m(\Sigma'). \tag{31}$$

§ 10. We can now proceed to examine the influence of the Earth's motion on optical phenomena in a system of transparent bodies. In discussing this problem we shall fix our attention on the variable electric moments in the particles or "atoms" of the system. To these moments we may apply what has been said in § 7. For the sake of simplicity we shall suppose that, in each particle, the charge is concentrated in a certain number of separate electrons, and that the "elastic" forces that act on one of these, and, conjointly with the electric forces, determine its motion, have their origin within the bounds of the *same* atom.

I shall show that, if we start from any given state of motion in a system without translation, we may deduce from it a corresponding state that can exist in the same system after a translation has been imparted to it, the kind of correspondence being as specified in what follows.

(a) Let A_1', A_2', A_3', etc., be the centres of the particles in the system without translation (Σ'); neglecting molecular motions we shall assume these points to remain at rest. The system of points A_1, A_2, A_3, etc., formed by the centres of the particles in the moving system Σ, is obtained from $A_1'A_2'A_3'$, etc., by means of a deformation $\left(\frac{1}{\beta l}, \frac{1}{l}, \frac{1}{l}\right)$. According to what has been said in § 8, the centres will of themselves take these positions A'_1, A'_2, A'_3 , etc., if originally, before there was a translation, they occupied the positions A_1, A_2, A_3, etc.

We may conceive any point P' in the space of the system Σ' to be displaced by the above deformation, so that a definite point P of Σ corresponds to it. For two corresponding points P' and P we shall define corresponding instants, the one belonging to P', the other to P, by stating that the true time at the first instant is equal to the local time, as determined by (5) for the point P, at the second instant. By corresponding times for two corresponding *particles* we shall understand times that may be said to correspond, if we fix our attention on the *centres* A' and A of these particles.

(b) As regards the interior state of the atoms, we shall assume that the configuration of a particle A in Σ at a certain time may be derived by means of the deformation $\left(\frac{1}{\beta l}, \frac{1}{l}, \frac{1}{l} \right)$ from the configuration of the corresponding particle in Σ', such as it is at the corresponding instant. In so far as this assumption relates to the form of the electrons themselves, it is implied in the first hypothesis of § 8.

Obviously, if we start from a state really existing in the system Σ', we have now completely defined a state of the moving system Σ. The question remains, however, whether this state will likewise be a possible one.

In order to judge of this, we may remark in the first place that the electric moments which we have supposed to exist in the moving system and which we shall denote by **P**, will be certain definite functions of the coordinates x, y, z of the centres A of the particles, or, as we shall say, of the coordinates of the particles themselves, and of the time t. The equations which express the relations between **P** on one hand and x, y, z, t on the other, may be replaced by other equations containing the vectors **P'** defined by (26) and the quantities x', y', z', t' defined by (4) and (5). Now, by the above assumptions a and b, if in a particle A of the moving system, whose coordinates are x, y, z, we find an electric moment **P** at the time t, or at the local time t', the vector **P'** given by (26) will be the moment which exists in the other system at the true time t' in a particle whose coordinates are x', y', z'. It appears in this way that the equations between $\mathbf{P}', x', y', z', t'$ are the same for both systems, the difference being only this, that for the system Σ' without translation these symbols indicate the moment, the coordinates, and the true time, whereas their meaning is different for the moving system, $\mathbf{P}', x', y', z, t'$ being here related to the moment **P**, the coordinates x, y, z and the general time t in the manner expressed by (26), (4), and (5).

It has already been stated that the equation (27) applies to both systems. The vector **D'** will therefore be the same in Σ' and Σ, provided we always compare corresponding places and times. However, this vector has not the same meaning in the two cases. In Σ' it represents the electric force, in Σ it is related to this force in the way expressed by (20). We may therefore conclude that the ponderomotive forces acting, in Σ and in Σ', on corresponding particles at corresponding instants, bear to each other the relation determined by (21). In virtue of our assumption (b), taken in connection with the second hypothesis of § 8, the same relation will exist between the "elastic" forces; consequently, the formula (21) may also be regarded as indicating the relation between the total forces, acting on corresponding electrons, at corresponding instants.

It is clear that the state we have supposed to exist in the moving system will really be possible if, in Σ and Σ', the products of the mass m and the acceleration of an electron are to each other in the same relation as the forces, i.e. if

$$m\mathbf{a}(\Sigma) = \left(l^2, \frac{l^2}{\beta}, \frac{l^2}{\beta}\right) m\mathbf{a}(\Sigma'). \tag{32}$$

Now, we have for the accelerations

$$\mathbf{a}(\Sigma) = \left(\frac{l}{\beta^3}, \frac{l}{\beta^2}\frac{l}{\beta 1}\right) \mathbf{a}(\Sigma') \tag{33}$$

as may be deduced from (4) and (5), and combining this with (32), we find for the masses

$$m(\Sigma) = (\beta^3 l, \beta l, \beta l) \, m(\Sigma').$$

If this is compared with (31), it appears that, whatever be the value of l, the condition is always satisfied, as regards the masses with which we have to reckon when we consider vibrations perpendicular to the translation. The only condition we have to impose on l is therefore

$$\frac{d(\beta l v)}{dv} = \beta^3 l.$$

But, on account of (3),

$$\frac{d(\beta v)}{dv} = \beta^3,$$

so that we must put

$$\frac{dl}{dv} = 0, \qquad l = \text{const.}$$

The value of the constant must be unity, because we know already that, for $v = 0, l = 1$.

We are therefore led to suppose *that the influence of a translation on the dimensions (of the separate electrons and of a ponderable body as a whole) is confined to those that have the direction of the motion, these becoming β times smaller than they are in the state of rest.* If this hypothesis is added to those we have already made, we may be sure that two states, the one in the moving system, the other in the same system while at rest, corresponding as stated above, may both be possible. Moreover, this correspondence is not limited to the electric moments of the particles. In corresponding points that are situated either in the ether between the particles, or in that surrounding the ponderable bodies, we shall find at corresponding times the same vector $\mathbf{D}'$ and, as is easily shown, the same vector $\mathbf{H}'$. We may sum up by saying: If, in the system without translation, there is a state of motion in which, at a definite place, the components of $\mathbf{P}$, $\mathbf{D}$, and $\mathbf{H}$ are certain functions of the time, then the same system after it has been put in motion (and thereby deformed) can be the seat of a state of motion in which, at the corresponding place, the components of $\mathbf{P}'$, $\mathbf{D}'$, and $\mathbf{H}'$ are the same functions of the local time.

There is one point which requires further consideration. The values of the masses m_1 and m_2 having been deduced from the theory of quasi-stationary motion, the question arises, whether we are justified in reckoning with them in the case of the rapid vibrations of light. Now it is found on closer examination that the motion of an electron may be treated as quasi-stationary if it changes very little during the time a light-wave takes to travel over a distance equal to the diameter. This condition is fulfilled in optical phenomena, because the diameter of an electron is extremely small in comparison with the wave-length.

§ 11. It is easily seen that the proposed theory can account for a large number of facts.

Let us take in the first place the case of a system without translation, in some parts of which we have continually $\mathbf{P} = 0, \mathbf{D} = 0, \mathbf{H} = 0$. Then, in the corresponding state for the moving system, we shall have in corresponding parts (or, as we may say, in the same parts of the deformed system) $\mathbf{P}' = 0, \mathbf{D}' = 0, \mathbf{H}' = 0$. These equations implying $\mathbf{P} = 0$, $\mathbf{D} = 0, \mathbf{H} = 0$, as is seen by (26) and (6), it appears that those parts which are dark while the system is at rest, will remain so after it has been put in motion. It will therefore be impossible to detect an influence of the Earth's motion on any optical experiment, made with a terrestrial source of light, in which the geometrical distribution of light and darkness is observed. Many experiments on interference and diffraction belong to this class.

In the second place, if, in two points of a system, rays of light of the same state of polarization are propagated in the same direction, the ratio between the amplitudes in these points may be shown not to be altered by a translation. The latter remark applies to those experiments in which the intensities in adjacent parts of the field of view are compared.

The above conclusions confirm the results which I formerly obtained by a similar train of reasoning, in which, however, the terms of the second order were neglected. They also contain an explanation of Michelson's negative result, more general than the one previously given, and of a somewhat different form; and they show why Rayleigh and Brace could find no signs of double refraction produced by the motion of the Earth.

As to the experiments of Trouton and Noble, their negative result becomes at once clear, if we admit the hypotheses of § 8. It may be inferred from these and from our last assumption (§ 10) that the only effect of the translation must have been a contraction of the whole system of electrons and other particles constituting the charged condenser and the beam and thread of the torsion-balance. Such a contraction does not give rise to a sensible change of direction.

It need hardly be said that the present theory is put forward with all due reserve. Though it seems to me that it can account for all well-established facts, it leads to some consequences that cannot as yet be put to the test of experiment. One of these is that the result of Michelson's experiment must remain negative, if the interfering rays of light are made to travel through some ponderable transparent body.

Our assumption about the contraction of the electrons cannot in itself be pronounced to be either plausible or inadmissible. What we know about the nature of electrons is very little, and the only means of pushing our way farther will be to test such hypotheses as I have here made. Of course, there will be difficulties, e.g. as soon as we come to consider the rotation of electrons. Perhaps we shall have to suppose that in those phenomena in which, if there is no translation, spherical electrons rotate about a diameter, the points of the electrons in the moving system will describe elliptic paths, corresponding, in the manner specified in § 10, to the circular paths described in the other case.

§ 12. There remain to be said a few words about molecular motion. We may conceive that bodies in which this has a sensible influence or even predominates, undergo the same deformation as the systems of particles of constant relative position of which alone we have spoken till now. Indeed, in two systems of molecules Σ' and Σ, the first without and the second with a translation, we may imagine molecular motions corresponding to each other in such a way that, if a particle in Σ' has a certain position at a definite instant, a particle in Σ occupies at the corresponding instant the corresponding position. This being assumed, we may use the relation (33) between the accelerations in all those cases in which the velocity of molecular motion is very small as compared with v. In these cases the molecular forces may be taken to be determined by the relative positions, independently of the velocities of molecular motion. If, finally, we suppose these forces to be limited to such small distances that, for particles acting on each other, the difference of local times may be neglected, one of the particles, together with those which lie in its sphere of attraction or repulsion, will form a system which undergoes the often mentioned deformation. In virtue of the second hypothesis of § 8 we may therefore apply to the resulting molecular force acting on a particle, the equation (21). Consequently, the proper relation between the forces and the accelerations will exist in the two cases, if we suppose *that the masses of all particles are influenced by a translation to the same degree as the electromagnetic masses of the electrons.*

§ 13. The values (30), which I have found for the longitudinal and transverse masses of an electron, expressed in terms of its velocity, are not the same as those that had been previously obtained by Abraham. The ground for this difference is to be sought solely in the circumstance that, in his theory, the electrons are treated as spheres of invariable dimensions. Now, as regards the transverse mass, the results of Abraham have been confirmed in a most remarkable way by Kaufmann's measurements of the deflection of radium-rays in electric and magnetic fields. Therefore, if there is not to be a most serious objection to the theory I have now proposed, it must be possible to show that those measurements agree with my values nearly as well as with those of Abraham.

I shall begin by discussing two of the series of measurements published by Kaufmann[13] in 1902. From each series he has deduced two quantities η and ζ the "reduced" electric and magnetic deflections, which are related as follows to the ratio $\gamma = v/c$:

$$\gamma = k_1 \frac{\zeta}{\eta}, \qquad \psi(\gamma) = \frac{\eta}{k_2 \zeta^2}. \qquad (34)$$

---

[13] Kaufmann, Phys. Zeitschr., 4, 1902, S. 55.

Here $\psi(\gamma)$ is such a function, that the transverse mass is given by

$$m_2 = \frac{3}{4} \frac{e^2}{6\pi c^2 R} \psi(\gamma), \tag{35}$$

whereas k_1 and k_2 are constant in each series.

It appears from the second of the formulae (30) that my theory leads likewise to an equation of the form (35); only Abraham's function $\psi(\gamma)$ must be replaced by

$$\frac{4}{3}\beta = \frac{4}{3}(1 - \gamma^2)^{-1/2}$$

Hence, my theory requires that, if we substitute this value for $\psi(\gamma)$ in (34), these equations shall still hold. Of course, in seeking to obtain a good agreement, we shall be justified in giving to k_1 and k_2 other values than those of Kaufmann, and in taking for every measurement a proper value of the velocity v, or of the ratio γ. Writing sk_1, $\frac{3}{4}k_2'$ and γ' for the new values, we may put (34) in the form

$$\gamma' = sk_1 \frac{\zeta}{\eta} \tag{36}$$

and

$$(1 - \gamma'^2)^{-1/2} = \frac{\eta}{k_2'\zeta^2}. \tag{37}$$

Kaufmann has tested his equations by choosing for k_1 such a value that, calculating γ and k_2 by means of (34), he obtained values for this latter number which, as well as might be, remained constant in each series. This constancy was the proof of a sufficient agreement.

I have followed a similar method, using, however, some of the numbers calculated by Kaufmann. I have computed for each measurement the value of the expression

$$k_2' = (1 - \gamma'^2)^{1/2} \psi(\gamma) k_2 \tag{38}$$

that may be got from (37) combined with the second of the equations (34). The values of $\psi(\gamma)$ and k_2 have been taken from Kaufmann's tables, and for γ' I have substituted the value he has found for γ, multiplied by s, the latter coefficient being chosen with a view to obtaining a good constancy of (38). The results are contained in the tables here, corresponding to the Tables III and IV in Kaufmann's paper.

The constancy of k_2' is seen to come out no less satisfactorily than that of k_2, the more so as in each case the value of s has been determined by means of only two measurements. The coefficient has been so chosen that for these two observations, which were in Table III the first and the last but one, and in Table IV the first and the last, the values of k_2' should be proportional to those of k_2 .

I shall next consider two series from a later publication by Kaufmann,[14] which have been calculated by Runge[15] by means of the method of least squares, the coefficients k_1 and k_2 having been determined in such a way that the values of η, calculated,

---

[14]Kaufmann, Gött. Nachr. Math.-phys. Klasse, 1903, S. 90.

[15]Runge, *ibid.*, p. 326.

for each observed ζ, from Kaufmann's equations (34), agree as closely as may be with the observed values of η.

I have determined by the same condition, likewise using the method of least squares, the constants a and b in the formula

$$\eta^2 = a\zeta^2 + b\zeta^4,$$

which may be deduced from my equations (36) and (37). Knowing a and b, I find γ for each measurement by means of the relation

$$\gamma = \sqrt{a}\,\frac{\zeta}{\eta}.$$

For two plates on which Kaufmann had measured the electric and magnetic deflections, the results are as follows, the deflections being given in centimetres.

I have not found time for calculating the other tables in Kaufmann's paper. As they begin, like the table for Plate 15 with a rather large negative difference between the values of η which have been deduced from the observations and calculated by Bunge, we may expect a satisfactory agreement with my formulae.

III. $s = 0,933.$

β	$\psi(\beta)$	k_2	β'	k_2'
0,851	2,147	1,721	0,794	2,246
0,766	1,86	1,736	0,715	2,258
0,727	1,78	1,725	0,678	2,256
0,6615	1,66	1,727	0,617	2,256
0,6075	1,595	1,655	0,567	2,175

IV. $s = 0,954.$

β	$\psi(\beta)$	k_2	β'	k_2'
0,963	3,23	8,12	0,919	10,36
0,949	2,86	7,99	0,905	9,70
0,933	2,73	7,46	0,890	9,28
0,883	2,31	8,32	0,842	10,36
0,860	2,195	8,09	0,820	10,15
0,830	2,06	8,13	0,792	10,23
0,801	1,96	8,13	0,764	10,28
0,777	1,89	8,04	0,741	10,20
0,752	1,83	8,02	0,717	10,22
0,732	1,785	7,97	0,698	10,18

Platte Nr. 15. $a = 0,06489$, $b = 0,3039$.

ζ	η					β	
	beobachtet	berechnet von R.	Diff.	berechnet von L.	Diff.	berechnet von	
						R.	L.
0,1495	0,0388	0,0404	− 16	0,0400	− 12	0,987	0,951
0,199	0,0548	0,0550	− 2	0,0552	− 4	0,964	0,918
0,2475	0,0716	0,0710	+ 6	0,0715	+ 1	0,930	0,881
0,296	0,0896	0,0887	+ 9	0,0895	+ 1	0,889	0,842
0,3435	0,1080	0,1081	− 1	0,1090	− 10	0,847	0,803
0,391	0,1290	0,1297	− 7	0,1305	− 15	0,804	0,763
0,437	0,1524	0,1527	− 3	0,1532	− 8	0,763	0,727
0,4825	0,1788	0,1777	+ 11	0,1777	+ 11	0,724	0,692
0,5265	0,2033	0,2039	− 6	0,2033	0	0,688	0,660

Platte Nr. 19. $a = 0,05867$, $b = 0,2591$.

ζ	η					β	
	beobachtet	berechnet von R.	Diff.	berechnet von L.	Diff.	berechnet von	
						R.	L.
0,1495	0,0404	0,0388	+ 16	0,0379	+ 25	0,990	0,954
0,199	0,0529	0,0527	+ 2	0,0522	+ 7	0,969	0,923
0,247	0,0678	0,0675	+ 3	0,0674	+ 4	0,939	0,888
0,296	0,0834	0,0842	− 8	0,0844	− 10	0,902	0,849
0,3435	0,1019	0,1022	− 3	0,1026	− 7	0,862	0,811
0,391	0,1219	0,1222	− 3	0,1226	− 7	0,822	0,773
0,437	0,1429	0,1434	− 5	0,1437	− 8	0,782	0,736
0,4825	0,1660	0,1665	− 5	0,1664	− 4	0,744	0,702
0,5265	0,1916	0,1906	+ 10	0,1902	+ 14	0,709	0,671

On the Dynamics of the Electron (5 June 1905)

H. Poincaré

Translation of H. Poincaré, Sur la dynamique de l'électron, *Compte rendus de l'Académie des Sciences*, t. 140, p. 1504-1508 (5 juin 1905)
Translated from French by André Michaud and Fritz Lewertoff

It seems at first glance that the aberration of light and related optical phenomena will provide us with a means of determining the absolute motion of the Earth, or rather its motion, not relative to other celestial bodies, but relative to the aether. This is not so; experiments in which only the first power of the aberration is taken into account failed at first, and the explanation was easily discovered; but Michelson, having devised an experiment in which the terms depending on the square of the aberration could be emphasized, was no happier. It seems that this impossibility of demonstrating absolute motion is a general law of nature.

An explanation was proposed by Lorentz, who introduced the hypothesis of a contraction of all bodies in the direction of the earth's motion; this contraction would account for Michelson's experiment and all those that have been carried out so far, but it would leave room for other experiments, even more delicate and easier to conceive than to carry out, which would be of such a nature as to show the absolute motion of the earth. However, if one considers the impossibility of such a finding as highly probable, it can be predicted that these experiments, if they are never carried out, will still give a negative result. Lorentz sought to complete and modify his hypothesis so as to bring it in line with the postulate of the *complete* impossibility of determining absolute motion. This he succeeded in doing in his article entitled *Electromagnetic phenomena in a system moving with any velocity smaller than that of light* (Proceedings of the Amsterdam Academy, May 27, 1904).

The importance of this issue determined me to take it up again; the results I have obtained agree on all the important points with those of Lorentz; I have only been led to modify and complete them in a few points of detail.

The essential point, established by Lorentz, is that the equations of the electromagnetic field are not altered by a certain transformation (which I will refer to as Lorentz's) and which is of the following form

$$x' = kl(x + \epsilon t), \qquad y' = ly, \qquad z' = lz, \qquad t' = kl(t + \epsilon x), \qquad (1)$$

x, y, z are the coordinates and t the time before the transformation, x', y', z' and t' after the transformation. Also ϵ is a constant that defines the transformation

$$k = \frac{1}{\sqrt{1 - \epsilon^2}}$$

and l is any function of ϵ. We can see that in this transformation the x-axis plays a particular role, but we can obviously construct a transformation where this role would be played by any line passing through the origin. The set of all these transformations, together with the set of all the rotations of space, must form a group; but, in order for this to be so, $l = 1$ is required; we are therefore led to suppose $l = 1$ and this is a consequence that Lorentz had obtained in another way. Let ρ be the electric density of the electron, ξ, η, ζ its velocity before the transformation; we will have for the same quantities $\rho', \xi', \eta', \zeta'$ after the transformation.

$$\rho' = \frac{k}{l^2}\rho(1 + \epsilon\xi), \qquad \rho'\xi' = \frac{k}{l^2}\rho(\xi + \epsilon). \qquad \rho'\eta' = \frac{\rho\eta}{l^3}, \qquad \rho'\zeta' = \frac{\rho\zeta}{l^23}. \qquad (2)$$

These formulas are slightly different from those found by Lorentz.

Let X, Y, Z and X', Y', Z' now be the three components of the force before and after the transformation, *the force is related to the unit of volume*; I find that

$$X' = \frac{k}{l^5}(X + \epsilon\Sigma X\xi), \qquad Y' = \frac{Y}{l^5}, \qquad Z' = \frac{Z}{l^5}, \qquad (3)$$

These formulas also differ slightly from those of Lorentz; the complementary term in $\Sigma X\xi$ is reminiscent of a result obtained in the past by Mr. Liénard.

If we now designate by X_1, Y_1, Z_1 and X_1', Y_1', Z_1', the components of the force related no longer to the unit of volume, but to the unit of mass of the electron, we will have

$$X_1' = \frac{k}{l^5}\frac{\rho}{\rho'}(X_1 + \epsilon\Sigma X_1\xi), \qquad Y_1' = \frac{\rho}{\rho'}\frac{Y_1}{l^5}, \qquad Z_1' = \frac{\rho}{\rho'}\frac{Z_1}{l^5}. \qquad (4)$$

Lorentz is also led to assume that the moving electron takes the form of a flattened ellipsoid; this is also the hypothesis made by Langevin, only, while Lorentz assumes that two of the axes of the ellipsoid remain constant, which is consistent with his hypothesis that $l = 1$, Langevin assumes that it is the volume that remains constant. Both authors have shown that these two hypotheses agree with Kaufmann's experiments, as well as with Abraham's primitive hypothesis (spherical electron). Langevin's hypothesis would have the advantage of being self-sufficient, since it is sufficient to look at the electron as being deformable and incompressible to explain that it takes on the ellipsoidal shape when in motion. But I show, agreeing in this with Lorentz, that it is unable to agree with the impossibility of an experiment showing absolute motion. This is because, as I said, $l = 1$ is the only hypothesis for which the set of Lorentz's transformations forms a group.

But with Lorentz's hypothesis, the agreement between the formulas is not achieved automatically; we obtain it, and at the same time a possible explanation of the contraction of *the electron, assuming that the electron, deformable and compressible, is subjected to a sort of constant external pressure whose work is proportional to the volume variations.*

I show, by applying the principle of least action, that, under these conditions, the compensation is complete, if we assume that inertia is an exclusively electromagnetic phenomenon, as has been generally accepted since Kaufmann's experiment, and that, apart from the constant pressure I just mentioned, which acts on the electron, all forces are of electromagnetic origin. This explains the impossibility of showing the absolute motion and the contraction of all bodies in the direction of the earth's motion.

But this is not all: Lorentz, in the cited work, deemed it necessary to complete his hypothesis by assuming that all forces, whatever their origin, are affected by a translation in the same manner as electromagnetic forces, and that, consequently, the effect produced on their components by the Lorentz transformation is still defined by equations (4).

It was important to examine this hypothesis more closely and in particular to find out what changes it would force us to make to the laws of gravitation.

This is what I tried to determine; I was first led to assume that the propagation of gravitation is not instantaneous, but occurs with the speed of light. This seems to contradict a result obtained by Laplace, which suggests that this propagation is, if not instantaneous, at least much faster than the speed of light. But, in reality, the question Laplace asked himself differs considerably from the one we are dealing with here. For Laplace, the introduction of a finite speed of propagation was the *only* change he made to Newton's law. Here, on the other hand, this modification is accompanied by several others; it is therefore possible, and in fact it happens, that a partial compensation occurs between them.

So, when we will speak of the position or velocity of the attracting body, it will be the position or velocity at the instant when the *gravitational wave* left that body; when we will speak of the position or velocity of the attracted body, it will be the position or velocity at the instant when that attracted body was reached by the gravitational wave emanating from the other body; it is clear that the first instant is prior to the second.

So if then x, y, z are the projections on the three axes of the vector that joins the two positions, if the velocity of the attracted body is ξ, η, ζ and that of the attracting body ξ_1, η_1, ζ_1 the three components of the attraction (which I could still call X_1, Y_1, Z_1 will be functions of $x, y, z, \xi, \eta, \zeta, \xi_1, \eta_1, \zeta_1$. I wondered if it was possible to determine these functions in such a way that they would be affected by the Lorentz transformation according to equations (4) and that the ordinary law of gravitation would be found, whenever the velocities $\xi, \eta, \zeta, \xi_1, \eta_1, \zeta_1$ are small enough that we can neglect the squares in front of the square of the speed of light.

The answer must be in the affirmative. We find that the corrected attraction is composed of two forces, one parallel to vector x, y, z, the other is the velocity ξ_1, η_1, ζ_1.

The divergence with the ordinary law of gravitation is, as I just said, of the order of ξ^2; if we only assumed, as Laplace did, that the speed of propagation is that of light, this divergence would be of the order of ξ, that is, 10,000 times greater. It is therefore not, at first glance, absurd to assume that astronomical observations are not precise enough to detect a divergence as small as the one we imagine. But this is what only a thorough discussion will allow to determine.

On the Dynamics of the Electron (23 July 1905)

H. Poincaré

Translation of H. Poincaré, Sur la dynamique de l'électron, *Rendiconti del Circolo matematico di Palermo*, **21** pp. 129-176 (1906)
Translated from French by André Michaud and Fritz Lewertoff. Typeset into LATEX by Dr. Reinoud Slagter.

Introduction

It seems at first glance that the aberration of light and the optical and electrical phenomena associated with it will provide us with a means of determining the absolute motion of the Earth, or rather its motion, not relative to other stars, but relative to the aether. Fresnel had already tried this, but he soon recognized that the motion of the Earth does not alter the laws of refraction and reflection. Analogous experiments, such as that of the telescope full of water and all those in which only the terms of the first order are taken into account in relation to the aberration, also gave only negative results; the explanation was soon discovered; but Michelson, having imagined an experiment in which the terms depending on the square of the aberration became sensitive, failed in his turn.

It seems that this impossibility of experimentally demonstrating the absolute motion of the Earth is a general law of Nature; we are naturally inclined to admit this law, which we will name the *Postulate of Relativity*, and to admit it without restriction. Whether this postulate, so far in agreement with experience, is to be confirmed or invalidated later on by more precise experiments, it is in any case interesting to see what the consequences may be.

An explanation has been proposed by Lorentz and Fitz Gérald, who introduced the hypothesis of a contraction undergone by all bodies in the direction of motion of the Earth and that would be proportional to the square of the aberration; this contraction, which we will call the *Lorentzian contraction*, would account for Michelson's experiment and all those carried out so far. This hypothesis would become insufficient, however, if one wanted to admit in all its generality the postulate of relativity.

Lorentz then sought to complete and modify it in order to bring it into perfect conformity with this postulate. This he succeeded in doing in his article entitled *Electromagnetic phenomena in a system moving with any velocity smaller than that of light* (Proceedings of the Amsterdam Academy, 27 May 1904).

The importance of this question has prompted me to take it up again; the results I have obtained agree with those of Mr Lorentz on all of the important issues; I have only been led to modify and supplement them on a few minor points; we will see later the differences, which are of secondary importance.

The idea of Lorentz can be summed up as follows: if we can, without modifying any of the apparent phenomena, give the whole system a common translation, it is because the equations of an electromagnetic medium are not altered by certain transformations, which we will call *Lorentz transformations*; two systems, one immobile, the other in translation, thus become the exact image of each other.

Langevin[1] did try to modify Lorentz's idea; for both authors, the moving electron takes the shape of a flattened ellipsoid, but for Lorentz, two of the axes of the ellipsoid remain constant, while for Langevin contrariwise, it is the volume of the ellipsoid that remains constant. Both scientists have moreover shown that these two hypotheses agree with Kaufmann's experiments, as well as with Abraham's initial hypothesis (non-deformable spherical electron). The advantage of Langevin's theory is that it only involves electromagnetic forces and binding forces; but it is incompatible with the postulate of relativity; this is what Lorentz had shown, this is what I find in turn in another way by appealing to the principles of group theory.

We must therefore come back to Lorentz's theory; but if we want to keep it and avoid intolerable contradictions, we must assume a special force that explains both the contraction and the constancy of two of the axes. I tried to determine this force and I found *that it can be assimilated to a constant external pressure, acting on the deformable and compressible electron, and whose work is proportional to the variations in the volume of this electron.*

If then the inertia of matter was exclusively of electromagnetic origin, as it is generally admitted since Kaufmann's experiment, and if, apart from this constant pressure that I just mentioned, all forces are of electromagnetic origin, then the postulate of relativity can be rigorously established. This is what I will demonstrate by a very simple calculation based on the least action principle.

But that's not all. Lorentz, in the work cited, considered it necessary to complete his hypothesis so that the postulate subsists even when there are forces other than electromagnetic forces. According to him, all forces, whatever their origin, are affected by the Lorentz's transformations (and therefore by a translation) in the same way as the electromagnetic forces.

It was therefore imperative to examine this hypothesis more closely and, in particular, to find out what changes it would require us to make to the laws of gravitation.

First of all, it forces us to assume that the propagation of gravitation is not instantaneous, but occurs at the speed of light. One might think that this is reason enough to reject the hypothesis, as Laplace demonstrated. But in reality, the effect of

---

[1]Langevin was preceded by Mr Bucherer from Bonn, who put forward the same idea before him (See: Bucherer, *Mathematische Einführung in die Elektronentheorie*, August 1904. Teubner, Leipzig).

this propagation is offset, for the most part, by a different cause, so that there is no longer any contradiction between the proposed law and astronomical observations.

Was it possible to find a law that satisfies the condition imposed by Lorentz, and that at the same time would reduce to Newton's law whenever the speeds of the stars are small enough so that their squares (as well as the product of accelerations by distances) can be neglected with respect to the square of the speed of Light?

To this question, as we will see below, the answer must be in the affirmative. Is the law as amended compatible with astronomical observations?

At face value, it would seem so, but the issue can only be resolved by means of a thorough analysis.

But even assuming that this analysis ends up favouring the new hypothesis, what should we conclude? If the propagation of the attraction occurs at the speed of light, this cannot be by chance, it must be because it is a property of the ether; and then we must try to penetrate the nature of this property, and link it to the other properties of the fluid.

We cannot be content with formulas that would simply be juxtaposed and that would only come together by chance; these formulas must interconnect, so to speak. The mind will not be satisfied unless it believes that it sees the reason for this agreement, to the point of having the illusion that it could have foreseen it.

But the issue can also be presented from another point of view, which will be better understood through a comparison. Let's suppose an astronomer before Copernicus and reflecting about the Ptolemy system; he will notice that for all planets, one of the two circles, epicycle or deferential, is travelled in the same time. This cannot be by chance, so there is between all the planets some mysterious connection.

But Copernicus, by simply changing the coordinate axes considered as fixed, makes this appearance disappear; each planet then describing only one circle, which causes the durations of the revolutions to become independent from each other (until Kepler re-establishes between them the link that was thought to have been destroyed).

It is possible that there is something similar here; if we admit the postulate of relativity, we would find in the law of gravitation and in the electromagnetic laws a common number which would be the speed of light; and we would find it again in all the other forces of any origin, which can only be explained in two ways:

Either there wouldn't be anything in the world that wasn't electromagnetic in origin.

Or this part that would be, so to speak, common to all physical phenomena would be only an appearance, something that would be due to our methods of measurement. How do we make our measurements? By carrying, one on top of the other, objects regarded as invariable solids, would we first answer; but this is no longer true in the current theory, if one admits the Lorentzian contraction. In this theory, two equal lengths are, by definition, two lengths that light takes the same time to travel.

Perhaps it would suffice to give up this definition for Lorentz's theory to be as completely overturned as was Ptolemy's system by the intervention of Copernicus. If this ever happens, this will not prove that the effort made by Lorentz was useless; for Ptolemy, whatever one may think of it, was not useless to Copernicus.

So I did not hesitate to publish these few partial results, although at this very

32

moment the whole theory may seem endangered by the discovery of magnetocathodic rays.

1. The Lorentz Transformation

Lorentz adopted a special system of units, so that the 4π factors disappeared in the formulas. I will do the same, and moreover I will choose the units of length and time in such a way that the speed of light will be equal to 1. Under these conditions the fundamental formulas become, calling f, g, h the electric displacement, α, β, γ the magnetic force, F, G, H the vector potential, ψ the scalar potential, ρ the electric density, ξ, η, ζ the speed of the electron, u, v, w the current:

$$\begin{cases} u = \dfrac{df}{dt} + \rho\xi = \dfrac{d\gamma}{dy} - \dfrac{d\beta}{dz}, & \alpha = \dfrac{dH}{dy} - \dfrac{dG}{dz}, & f = -\dfrac{df}{dt} - \dfrac{d\psi}{dx}, \\ \dfrac{d\alpha}{dt} = \dfrac{dg}{dz} - \dfrac{dh}{dy}, & \dfrac{d\rho}{dt} = \sum \dfrac{d\rho\xi}{dx} = 0, & \sum \dfrac{df}{dx} = \rho, \quad \dfrac{d\psi}{dt} + \sum \dfrac{dF}{dx} = 0, \quad (5) \\ \Box \equiv \Delta - \dfrac{d^2}{dt^2} = \sum \dfrac{d^2}{dx^2} - \dfrac{d^2}{dt^2}, & \Box\psi = -\rho, \quad \Box F = -\rho\xi. \end{cases}$$

An element of matter of volume $dxdydz$, undergoes a mechanical force whose components $Xdxdydz, Ydxdydz, Zdxdydz$ are deduced from the formula:

$$X = \rho f + \rho(\eta\gamma - \zeta\beta). \quad (6)$$

These equations are susceptible to a remarkable transformation discovered by Lorentz and that owes its interest to the fact that it explains why no experiment is likely to make us know the absolute motion of the universe. Let us pose:

$$x' = kl(x + \epsilon t), \quad t' = kl(t + \epsilon x), \quad y' = ly, \quad z' = lz, \quad (7)$$

l and ϵ being any two constants, and given that

$$k = \frac{1}{\sqrt{1 - \epsilon^2}}.$$

If then we pose:

$$\Box' = \sum \frac{d^2}{dx'^2} - \frac{d^2}{dt'^2},$$

we obtain:

$$\Box' = \frac{1}{l^2}\Box.$$

Let us consider a sphere dragged with the electron in a uniform translational motion and given:

$$(x - \xi t)^2 + (y - \eta t)^2 + (z - \zeta t)^2 = r^2,$$

then the equation of this mobile sphere whose volume will be $\frac{4}{3}\pi r^2$ The transformation will convert it into an ellipsoid, whose equation is easy to find. Indeed, we can easily deduce from equation (3):

$$x = \frac{k}{l}(x' - \epsilon t'), \quad t = \frac{k}{l}(t' - \epsilon x'), \quad y = \frac{y'}{l}, \quad z = \frac{z'}{l}, \qquad \text{(7 bis)}$$

The ellipsoid equation then becomes:

$$k^2(x' - \epsilon t' + \xi t' + \epsilon \xi x')^2 +)y' - \eta k t' + \eta k \epsilon x')^2 + (z' - \zeta k t' + \zeta k \epsilon x')^2 = l^2 r^2.$$

This ellipsoid is animated by a uniform motion, for $t' = 0$, it reduces to

$$k^2 x'^2(1 - \xi \epsilon)^2 +)y' + \eta k \epsilon x')^2 + (z' + \zeta k \epsilon x')^2 = l^2 r^2$$

and has for volume:

$$\frac{4}{3}\pi r^3 \frac{l^3}{k(1 + \epsilon \xi)}$$

If we want the charge of an electron not to be altered by the transformation, and if we name ρ' the new electrical density, we obtain:

$$\rho' = \frac{k}{l^3}(\rho + \epsilon \rho \xi). \qquad \text{(8)}$$

What will the new velocities ξ', η', ζ' now be? we will need to have:

$$\xi' = \frac{dx'}{dt'} = \frac{d(x + \epsilon t)}{d(t + \epsilon x)} = \frac{\xi + \epsilon}{1 + \epsilon \xi}, \quad \eta' = \frac{dy'}{dt'} = \frac{dy}{kd(t + \epsilon x)} = \frac{\eta}{k(1 + \epsilon \xi)}, \quad \zeta' = \frac{\zeta}{k(1 + \epsilon \xi)}$$

whence:

$$\rho'\xi' = \frac{k}{l^3}(\rho \xi + \epsilon \rho), \quad \rho'\eta' = \frac{1}{l^3}\rho \eta, \quad \rho'\zeta' = \frac{1}{l^3}\rho \zeta \qquad \text{(8 bis)}$$

It is here that I must point out for the first time a discrepancy with Lorentz. Lorentz poses (with the exception of the notations) (loc. Cit., page 813, formulas 7 and 8):

$$\rho' = \frac{1}{kl^3}\rho, \quad \xi' = k^2(\xi + \epsilon), \quad \eta' = k\eta, \quad \zeta' = k\zeta..$$

We thus recover formulas:

$$\rho'\xi' = \frac{k}{l^3}(\rho \xi + \epsilon \rho), \quad \rho'\eta' = \frac{1}{l^3}\rho \eta, \quad \rho'\zeta' = \frac{1}{l^3}\rho \zeta$$

but the value of ρ' differs. It is important to note that equation (4) satisfy the continuity condition

$$\frac{d\rho'}{dt'} + \sum \frac{d\rho'\xi'}{dx'} = 0.$$

Given an indeterminate quantity λ and D the functional determinant of

$$t + \lambda \rho, \quad x + \lambda \rho \xi, \quad y + \lambda \rho \eta, \quad z + \lambda \rho \zeta, \qquad \text{(9)}$$

with respect to t, x, y, z. We will have:

$$D = D_0 + D_1\lambda + D_2\lambda^2 + D_3\lambda^3 + D_4\lambda^4,$$

with

$$D_0 = 1, \quad D_1 = \frac{d\rho}{dt} + \sum \frac{d\rho\xi}{dx} = 0.$$

Given $\lambda' = l^2\lambda$, we see that the 4 functions

$$t' + \lambda'\rho', \quad x' + \lambda'\rho'\xi', \quad y' + \lambda'\rho'\eta', \quad z' + \lambda'\rho'\zeta' \qquad (9 \text{ bis})$$

are linked to functions (5) by the same linear relationships as the old variables are linked to the new variables. So if we designate by D' the functional determinant of functions (5) with respect to the new variables, we will have:

$$D' = D, \quad D' = D'_0 + D'_1\lambda' + \ldots D'_4\lambda'^4$$

whence:

$$D'_0 = D_0 = 1, \quad D'_1 - \frac{D_1}{l^2} = 0 = \frac{d\rho'}{dt'} + \sum \frac{d\rho'\xi'}{dx'}.$$

Q.E.D.

With Lorentz's hypothesis, this condition would not be fulfilled, since ρ' does not have the same value. We will define the new potentials, vector and scalar, in order to satisfy conditions

$$\Box'\psi' = -\rho', \quad \Box'F' == \rho'\xi' \qquad (10)$$

From which we will draw:

$$\psi' = \frac{k}{l}(\psi + \epsilon F), \quad F' = \frac{k}{l}(F + \epsilon\psi), \quad G' = \frac{1}{l}G, \quad H' = \frac{1}{l}H \qquad (11)$$

These formulas differ significantly from those of Lorentz, but in final analysis the discrepancy relates only to the definitions. We will choose the new electric and magnetic fields in order to satisfy equations:

$$f' = -\frac{dF'}{dt'} - \frac{d\psi'}{dx'}, \quad \alpha' = -\frac{dH'}{dy'} - \frac{dG'}{dz'} \qquad (12)$$

It is easy to see that:

$$\frac{d}{dt'} = \frac{k}{l}\left(\frac{d}{dt} - \epsilon\frac{d}{dx}\right), \quad \frac{d}{dx'} = \frac{k}{l}\left(\frac{d}{dx} - \epsilon\frac{d}{dt}\right), \quad , \frac{d}{dy'} = \frac{1}{l}\frac{d}{dy}, \quad \frac{d}{dz'} = \frac{1}{l}\frac{d}{dz}$$

and we conclude:

$$\begin{cases} f' = \frac{1}{l^2}f, \quad g' = \frac{k}{l^2}(g + \epsilon\gamma), \quad h' = \frac{k}{l^2}(h - \epsilon\beta), \\ \alpha' = \frac{1}{l^2}\alpha, \quad \beta' = \frac{k}{l^2}(\beta - \epsilon h), \quad \gamma' = \frac{k}{l^2}(\gamma + \epsilon g). \end{cases} \qquad (13)$$

These formulas are identical to those of Lorentz. Our transformation does not alter equations (1). Indeed, the continuity condition, as well as equations (6) and (8), already provide us with some of equations (1) (except for the accented letters). Equations (6) related to the continuity condition give:

$$\frac{d\psi'}{dt'} + \sum \frac{dF'}{dx'} = 0. \tag{14}$$

What remains to be established is:

$$\frac{df'}{dt'} + \rho'\xi' = \frac{d\gamma'}{dy'} - \frac{d\beta'}{dz'}, \quad \frac{dz'}{dt'} = \frac{dg'}{dz'} - \frac{dh'}{dy'}, \quad \sum \frac{df'}{dx'} = \rho'$$

and it is easy to see that these are necessary consequences of equations (6), (8) and (10). We must now compare the forces before and after the transformation. Let X, Y, Z be the force before, and X', Y', Z' be the force after the transformation, both referring to the unit of volume. For X' to satisfy the same equations as before the transformation, we must have:

$$X' = \rho' f' + \rho'(\eta'\gamma' - \zeta'\beta'),$$
$$y' = \rho' g' + \rho'(\zeta'\alpha' - \xi'\gamma'),$$
$$Z' = \rho' h' + \rho'(\xi'\beta' - \eta'\alpha'),$$

where, by replacing all quantities by their values (4), (4bis) and (9) and taking into account equations (2):

$$\begin{cases} X' = \frac{k}{l^5}(X + \epsilon \sum X\xi) \\ Y' = \frac{1}{l^5}Y \\ Z' = \frac{1}{l^5}Z. \end{cases} \tag{15}$$

If we represented by X_1, Y_1, Z_1 the components of the force, no longer with respect to the unit of volume, but with respect to the unit of electric charge of the electron, and by X_1', Y_1', Z_1' the same quantities after the transformation, we would have:

$$X_1 = f + \eta\gamma - zeta\beta, \quad X_1' = f' + \eta'\gamma' - \zeta'\beta', \quad X = \rho X_1, \quad X' = \rho' X_1'$$

and we would have equations:

$$\begin{cases} X_1' = \frac{k}{l^5}\frac{\rho}{\rho'}(X_1 + \epsilon \sum X_1\xi \\ Y_1' = \frac{1}{l^5}\frac{\rho}{\rho'}Y_1 \\ Z_1' = \frac{1}{l^5}\frac{\rho}{Z'}Y_1. \end{cases} \tag{15 bis}$$

Lorentz had found [with the exception of the notations, page 813, formula (10)]:

$$\begin{cases} X_1' = l^2 X_1' - l^2\epsilon(\eta'g' + \zeta'h') \\ Y_1' = \frac{l^2}{k}Y_1' + \frac{l^2\epsilon}{k}\xi'g' \\ Z_1' = \frac{l^2}{k}Z_1' + \frac{l^2\epsilon}{k}\xi'h'. \end{cases} \tag{15 ter}$$

Before going any further, it is important to investigate the cause of this important discrepancy. It is obviously due to the fact that the formulas for ξ', η', ζ' are not the same, while the formulas for electric and magnetic fields are the same.

If the inertia of the electrons is exclusively of electromagnetic origin and if more-over they are only subjected to forces of electromagnetic origin, the condition of equilibrium requires that we have inside the electrons:

$$X = Y = Z = 0.$$

However, according to equations (11) these relationships are equivalent to

$$X' = Y' = Z' = 0.$$

The equilibrium conditions of the electrons are therefore not altered by the transformation.

Unfortunately such a simple hypothesis is inadmissible. Indeed, if we assume $\xi = \eta = \zeta = 0$, the conditions $X = Y = Z = 0$ would result in $f = g = h = 0$, and therefore $\sum \frac{df}{dx} = 0$ which means that $\rho = 0$. Similar results would be obtained in the most general case. So we have to admit that there are either other forces or bounds in addition to the electromagnetic forces. It is then necessary to find out what conditions these forces or bounds must satisfy, so that the equilibrium of electrons is not disturbed by the transformation. This will be the subject of a later paragraph.

2. The Principe of Least Action

We know how Lorentz deduced his equations from the principle of least action. I will, however, come back to this issue, although I have nothing essential to add to Lorentz's analysis, because I prefer to present it in a slightly different form that will be useful for my purpose. I will pose:

$$J = \int dt\, d\tau \left[\frac{1}{2} \sum f^2 + \frac{1}{2} \sum \alpha^2 - \sum Fu \right], \tag{1}$$

assuming that f, α, F, u, etc. are subject to the following conditions and to those that would be inferred symmetrically:

$$\sum \frac{df}{dx} = \rho, \quad \alpha = \frac{dH}{dy} - \frac{dG}{dz}, \quad u = \frac{df}{dt} + \rho\xi. \tag{2}$$

As for the integral J, it must be extended:

1⁰ with respect to volume element $dr = dxdydz$ to the entire space;

2⁰ with respect to time t, to the interval between limits $t = t_0, t = t_1$.

According to the principle of least action, integral J must be a minimum, if the various quantities contained therein are subjected:

1⁰ to conditions (2);

2⁰ to the condition that the state of the system is determined at the two limit times $t = t_0, t = t_1$.

This last condition allows us to transform our integrals by integration by parts over time. If we indeed have an integral of the form

$$\int dt\, d\tau\, A \frac{dB\delta C}{dt},$$

where C is one of the quantities that define the state of the system and δC its variation, it will be equal to (integrating by parts with respect to time):

$$\int d\tau |AB\, \delta C|_{t=t_0}^{t=t_1} - \int dt\, d\tau \frac{dA}{dt} dB\delta C.$$

Since the state of the system is determined at the two boundary time limits, we have $\delta C = 0$ for $t = t_0, t = t_1$; thus the first integral related to these two boundaries is null, and only the second remains. Similarly, we can also integrate by parts with respect to x, y or z; we indeed have

$$\int A \frac{dB}{dx} dx\, dy\, dz\, dt = \int A\, dy\, dz\, dt - \int B \frac{dA}{dx} dx\, dy\, dz\, dt.$$

Since our integrations extend to infinity, we have to do $x = \pm\infty$ in the first integral of the second member; therefore, since we always assume that all our functions cancel each other out infinitely, this integral will be null and void and we will obtain

$$\int A \frac{dB}{dx} d\tau\, dt = - \int B \frac{dA}{dx} d\tau\, dt.$$

If the system were supposed to be subject to bounding, these bounding conditions would have to be added to the conditions imposed on the various quantities in integral J.

Let's first give to F, G, H the increasements $\delta F, \delta G, \delta H$, where:

$$\delta\alpha = \frac{d\delta H}{dy} - \frac{d\delta G}{dz}.$$

We should have

$$\delta J = \int dt\, d\tau \left[\sum \left(\delta G \frac{d\alpha}{dz} - \delta H \frac{d\alpha}{dy} \right) - \sum u\delta F \right] = 0,$$

Or, integrating by parts,

$$\delta J = \int dt\, d\tau \left[\sum \left(\delta G \frac{d\alpha}{dz} - \delta H \frac{d\alpha}{dy} \right) - \sum u\delta F \right] = - \int dt\, d\tau \sum \delta F \left(u - \frac{d\gamma}{dy} + \frac{d\beta}{dz} \right) = 0,$$

Hence, by setting the coefficient of arbitrary value δF to zero,

$$u = \frac{d\gamma}{dy} - \frac{d\beta}{dz}. \tag{3}$$

This relationship gives us (with an integration by parts):

$$\int \sum F u \, d\tau = \int \sum F\left(\frac{d\gamma}{dy} - \frac{d\beta}{dz}\right) d\tau = \int \sum \left(\beta\frac{dF}{dz} - \gamma\frac{dF}{dz} - \gamma\frac{dF}{dy}\right) d\tau$$
$$= \int \sum \alpha\left(\frac{dH}{dy} - \frac{dG}{dz}\right) d\tau,$$

or

$$\int \sum F u \, d\tau = \int \sum \alpha^2 d\tau,$$

hence, finally:

$$J = \int dt \, d\tau \left(\frac{\sum f^2}{2} - \frac{\sum \alpha^2}{2}\right). \tag{4}$$

From now on, and thanks to relation (3), δJ is independent of δF and therefore independent of $\delta \alpha$; let us now vary the other variables. We obtain, by returning to expression (1) of J,

$$\delta J = \int dt \, d\tau \left(\sum f \delta f - \sum F \delta u\right).$$

But f, g, h are subject to the first of conditions (2), so that

$$\sum \frac{d\delta f}{dx} = \delta\rho \tag{5}$$

and we can write:

$$\delta J = \int dt \, d\tau \left[\sum f df - \sum F \delta u - \psi\left(\sum \frac{d\delta f}{dx}\delta\rho\right)\right]. \tag{6}$$

The principles of variation calculation teach us that the calculation must be done as if, ψ being an arbitrary function, δJ was represented by expression (6) and if the variations were no longer subject to condition (5). On the other hand, we have

$$\delta u = \frac{d\delta f}{dt} + \delta\rho\xi,$$

whence, after integration in parts,

$$\delta J = \int dt \delta\tau \sum \delta f\left(f + \frac{dF}{dt} + \frac{d\psi}{dx}\right) + \int dt d\tau \left(\psi\delta\rho - \sum F\delta\rho\xi\right). \tag{7}$$

If we first assume that electrons do not undergo any variation, then $\delta J = 0$ and the second integral is zero. Since δJ must nullify, we must have:

$$f + \frac{dF}{dt} + \frac{d\psi}{dx} = 0. \tag{8}$$

Therefore, it remains in the general case:

$$\delta J = \int dt d\tau \left(\psi\delta\rho - \sum F\delta\rho\xi\right). \tag{9}$$

The forces acting on the electrons have yet to be determined. For this we must assume that to each electron element a complementary force $-X dr, -Y dr, -Z dr$, is applied, and write that this force balances the forces of electromagnetic origin. Let U, V, W be the components of the displacement of the electron element dr, displacement counted from any initial position. Let $\delta U, \delta V, \delta W$ be the variations of this displacement; the corresponding virtual work of the complementary force will be:

$$-\int \sum X \delta U d\tau,$$

so that the equilibrium condition we've just talked about will write:

$$\delta J = -\int \sum X \delta U d\tau dt. \tag{10}$$

The idea is about transforming δJ. To do this, let's start by looking for the continuity equation expressing that the charge of an electron is conserved through the variation. Let $x0, y0, z0$ be the initial position of an electron. Its current position will be:

$$x = x_0 + U, \quad y = y_0 + V, \quad z = z_0 + W.$$

In addition, we will introduce an auxiliary variable ϵ, that will produce the variations of our various functions, so that, for any function A, we will have:

$$\delta A = \delta\epsilon \frac{dA}{d\epsilon}.$$

It will indeed be convenient for me to be able to switch from the notation of the variation calculation to that of the ordinary differential calculation, or vice versa.

Our functions could then be seen: 1° either as depending on the five variables x, y, z, t, ϵ, so that we always remain in the same place when only ϵ and t vary: we will then designate their derivatives by ordinary d; 2° or as depending on the five variables $x_0, y_0, z_0, t, \epsilon$, so that we always follow the same electron when only t and ϵ vary; we will then designate their derivatives by ∂. We will then have:

$$\xi = \frac{\partial U}{\partial t} = \frac{dU}{dt} + \xi\frac{dU}{dx} + \eta\frac{dU}{dy} + \zeta\frac{dU}{dz} = \frac{\partial x}{\partial t}. \tag{11}$$

Let us now designate by Δ the functional determinant of x, y, z with respect to x_0, y_0, z_0:

$$\Delta = \frac{\partial(x, y, z)}{\partial(x_0, y_0, z_0)}.$$

If ϵ, x_0, y_0, z_0 remain constant, and if we give t an increase ∂t, then for x, y, z there will be increases $\partial x_0, \partial y_0, \partial z_0$, and for Δ an increase $\partial \Delta$, and we will have:

$$\partial x = \xi \partial t, \quad \partial y = \eta \partial t, \quad \partial z = \zeta \partial t$$
$$\Delta + \partial \Delta = \frac{(x + \partial x, y + \partial y, z + \partial z)}{\partial(x_0, y_0, z_0)},$$

whence

$$1 + \frac{\partial \Delta}{\Delta} = \frac{\partial(x, y + \partial y, z + \partial z)}{\partial(x, y, z)} = \frac{\partial(x + \xi \partial t, y + \eta \partial t, z + \zeta \partial t)}{\partial(x, y, z)}.$$

From which we infer:

$$\frac{1}{\Delta} \frac{\partial \Delta}{\partial t} = \frac{d\xi}{dx} + \frac{d\eta}{dy} + \frac{d\zeta}{P} dz. \tag{12}$$

The mass of each electron being invariant, we will have:

$$\frac{\partial \rho \Delta}{\partial t} = 0, \tag{13}$$

whence

$$\frac{\partial \rho}{\partial t} + \sum \rho \frac{d\xi}{dx} = 0, \quad \frac{\partial \rho}{\partial t} + \frac{\partial \rho}{dt} + \sum \xi \frac{d\rho}{dx}, \quad \frac{d\rho}{dt} + \sum \frac{d\rho \xi}{dx} = 0.$$

These are the different forms of the continuity equation with respect to variable t. We find similar forms with respect to variable ϵ. So let:

$$\delta U = \frac{\partial U}{\partial \epsilon}, \quad \partial V = \frac{\partial V}{\partial \epsilon} \delta \epsilon, \quad \delta W = \frac{\partial W}{\partial \epsilon} \delta \epsilon.$$

We will obtain:

$$\delta U = \frac{dU}{d\epsilon} \delta \epsilon + \delta U \frac{dU}{dx} + \delta V = \frac{dU}{dy} + \delta W \frac{dU}{dz}, \tag{11 bis}$$

$$\frac{1}{\Delta} \frac{\partial \Delta}{\partial \epsilon} = \sum \frac{dU}{d\epsilon}, \quad \frac{\partial \rho \Delta}{\partial \epsilon} = 0, \tag{12 bis}$$

$$\delta \epsilon \frac{\partial \rho}{\partial \epsilon} + \sum \rho \frac{d\rho U}{dx} = 0, \quad \frac{\partial \rho}{\partial \epsilon} = \frac{d\rho}{d\epsilon} + \sum \frac{\delta U}{\delta \epsilon} \frac{d\rho}{dx}, \quad \delta \rho + \frac{d\rho \, \delta U}{dx} = 0. \tag{13 bis}$$

We will note the difference between the definition of $\delta U = \frac{\partial U}{\partial \epsilon} \delta \epsilon$ and that of $\delta \rho = \frac{d\rho}{d\epsilon} \delta \epsilon$, we note that it is indeed this definition of δU that fits formula (10).

This last equation will allow us to transform the first term of (9); we find indeed:

$$\int dt d\tau \psi \delta \rho = - \int dt d\tau \psi \sum \frac{d\rho \delta U}{dx},$$

or, integrating in parts,

$$\int dt d\tau \psi \delta \rho = \int dt d\tau \sum \rho \frac{d\psi}{dx} \delta U. \tag{14}$$

Let us now proceed to determine

$$\delta(\rho \xi) = \frac{d(\rho \xi)}{d\epsilon} \delta \epsilon.$$

Let us observe that $\rho \Delta$ can only depend on x_0, y_0, z_0; indeed, if we consider an electron element whose initial position is a rectangular parallelepiped whose edges are dx_0, dy_0, dz_0 the charge of this element is

$$\rho \Delta dx_0 dy_0 dz_0$$

and, this load having to remain constant, we have:

$$\frac{\partial \rho \Delta}{\partial t} = \frac{\partial \rho \Delta}{\partial \epsilon} = 0. \tag{15}$$

From which we infer:

$$\frac{\partial^2 \rho \Delta U}{\partial t \partial \epsilon} = \frac{\partial}{\partial \epsilon}\left(\rho \Delta \frac{\partial U}{\partial t}\right) = \frac{\partial}{\partial t}\left(\rho \Delta \frac{\partial U}{\partial \epsilon}\right) \tag{16}$$

Now we know that for any function A we have, by the continuity equation,

$$\frac{1}{\Delta}\frac{\partial A\Delta}{\partial t} = \frac{dA}{dt} + \sum \frac{dA\xi}{dx}$$

and, similarly

$$\frac{1}{\Delta}\frac{\partial A\Delta}{\partial \epsilon} = \frac{dA}{d\epsilon} + \sum \frac{dA\frac{\partial U}{\partial \epsilon}}{dx}$$

So we have:

$$\frac{1}{\Delta}\frac{\partial}{\partial \epsilon}\left(\rho \Delta \frac{\partial U}{\partial t}\right) = \frac{d\rho\frac{\partial U}{\partial t}}{d\epsilon} + \frac{d\left(\rho\frac{\partial U}{\partial t}\frac{\partial U}{\partial \epsilon}\right)}{dx} + \frac{d\left(\rho\frac{\partial U}{\partial t}\frac{\partial V}{\partial \epsilon}\right)}{dy} + \frac{d\left(\rho\frac{\partial U}{\partial t}\frac{\partial W}{\partial \epsilon}\right)}{dz}, \tag{17}$$

$$\frac{1}{\Delta}\frac{\partial}{\partial t}\left(\rho \Delta \frac{\partial U}{\partial \epsilon}\right) = \frac{d\rho\frac{\partial U}{\partial \epsilon}}{dt} + \frac{d\left(\rho\frac{\partial U}{\partial t}\frac{\partial U}{\partial \epsilon}\right)}{dx} + \frac{d\left(\rho\frac{\partial V}{\partial t}\frac{\partial U}{\partial \epsilon}\right)}{dy} + \frac{d\left(\rho\frac{\partial W}{\partial t}\frac{\partial U}{\partial \epsilon}\right)}{dz}, \tag{17 bis}$$

The second members of (17) and (17bis) must be equal, and if we remember that

$$\frac{\partial U}{\partial t} = \xi, \quad \frac{\partial U}{\partial \epsilon} = \delta U, \quad \frac{d\rho\xi}{d\epsilon}\delta\epsilon = \delta\rho\xi$$

It follows that:

$$\delta\rho\xi + \frac{d(\rho\xi\delta U)}{dx} + \frac{d(\rho\xi\delta V)}{dy} + \frac{d(\rho\xi\delta W)}{dz} = \frac{d(\rho\delta U)}{dt} + \frac{d(\rho\xi\delta U)}{dx} + \frac{d(\rho\eta\delta U)}{dy} + \frac{d(\rho\zeta\delta U)}{dz} \tag{18}$$

Let us now transform the second term of (9); it follows that:

$$\int dt d\tau \sum F\delta\rho\xi = \int dt d\tau \Big[\sum F\frac{d(\rho\delta U)}{dt} + \sum F\frac{d(\rho\eta\delta U)}{dy}$$
$$+ \sum F\frac{d(\rho\zeta\delta U)}{dz} - \sum F\frac{d(\rho\xi\delta V)}{dy} - \sum F\frac{d(\rho\xi\delta W)}{dz}\Big].$$

The second member becomes, through integration by parts:

$$\int dt d\tau\Big[-\sum \rho\delta U\frac{dF}{dt} - \sum \rho\eta\delta U\frac{dF}{dy} - \sum \rho\zeta\delta U\frac{dF}{dz} + \sum \rho\xi\delta V\frac{dF}{dy} + \sum \rho\xi\delta W\frac{dF}{dz}\Big].$$

Let us note now that:

$$\sum \rho\xi\delta V\frac{dF}{dy} = \sum \rho\zeta\delta U\frac{dH}{dx}, \quad \sum \rho\xi\delta W\frac{dF}{dz} = \sum \rho\eta\delta U\frac{dG}{dx}.$$

If, in fact, in both members of these relationships, the ?s are developed, they become identities; and let us remember that

$$\frac{dH}{dx} - \frac{dF}{dz} = -\beta, \quad \frac{dG}{dx} - \frac{dF}{dy} = \gamma,$$

the second member in question will become:

$$\int dt d\tau \left[-\sum \rho \delta U \frac{dF}{dt} - \sum \rho \gamma \eta \delta U - \sum \rho \beta \zeta \delta U \right],$$

so that finally:

$$\delta J = \int dt d\tau \sum \rho \delta U \left(\frac{\psi}{dx} + \frac{dF}{dt} + \beta \zeta - \gamma \eta \right) = \int dt d\tau \sum \rho \delta U (-f + \beta \zeta - \gamma \eta).$$

By equating the coefficient of ?U in both members of (10), we obtain:

$$X = f - \beta \zeta + \gamma \eta.$$

And this is equation (2) of the previous paragraph.

3. The Lorentz Transformation and the Principle of Least Action

Let's see if the principle of least action gives us the reason for the success of the Lorentz transformation. First we must see what this transformation does to the integral:

$$J = \int dt d\tau \left(\frac{\sum f^2}{2} - \frac{\sum \alpha^2}{2} \right).$$

[formula (4) of paragraph 2]. First, we find

$$dt' d\tau' = l^4 \, dt \, d\tau,$$

because x', y', z', t' are linked to x, y, z, t by linear relations whose determinant is equal to l^4, it follows that:

$$\begin{cases} l^4 \sum f'^2 = f^2 + k^2(g^2 + h^2) + k^2 \epsilon^2 (\beta^2 + \gamma^2) + 2k^2(g\gamma - h\beta) \\ l^4 \sum \alpha'^2 = \alpha^2 + k^2(\alpha^2 + \gamma^2) + k^2 \epsilon^2 (g^2 + h^2) + 2k^2(g\gamma - h\beta) \end{cases} \tag{1}$$

[formulas (9) from paragraph 1], whence:

$$l^4 \left(\sum f'^2 - \sum \alpha'^2 \right) = \sum f^2 - \sum \alpha^2,$$

so that if we pose:

$$J' = \int dt' d\tau' \left(\frac{\sum f'^2}{2} - \frac{\sum \alpha'^2}{2} \right),$$

We obtain:
$$J' = J.$$

For this equality to be justified, however, the limits of integration must be the same; so far we have accepted that t varies from t_0 to t_1, and x, y, z from $-\infty$ to $+\infty$. At this point the integration limits would be altered by the Lorentz transformation; but nothing prevents us from assuming $t_0 = -\infty, t_1 = +\infty$. With these conditions the limits are the same for J and for J'.

We then have to compare the following two equations analogous to equation (10) in paragraph 2:

$$\begin{cases} \delta J = - \int \sum X \delta U d\tau dt \\ \delta J' = - \int \sum X' \delta U' d\tau' dt'. \end{cases} \quad (2)$$

To accomplish this, first we must compare $\delta U'$ with δU. Let's consider an electron whose initial coordinates are x_0, y_0, z_0; its coordinates at instant t, will be

$$x = x_0 + U, \quad y = y_0 + V, \quad z = z_0 + W.$$

If we consider the corresponding electron after the Lorentz transformation, it will have the following coordinates

$$x' = kl(x + \epsilon t), \quad y' = ly, \quad z' = lz,$$

where

$$x' = x_0 + U', \quad y' = y_0 + V', \quad z' = z_0 + W',$$

but he won't reach those coordinates until the instant

$$t' = kl(t + \epsilon x).$$

If we cause our variables to vary $\delta U, \delta V, \delta W$ and at the same time give t an increase δt, the coordinates x, y, z will undergo a total increase

$$\delta x = \delta U + \xi \delta t, \quad \delta y = \delta V + \eta \delta t, \quad \delta z = \delta W + \zeta \delta t.$$

Similarly, we will have:

$$\delta x' = \delta U' + \xi' \delta t', \quad \delta y' = \delta V' + \eta' \delta t', \quad \delta z' = \delta W' + \zeta \delta t'.$$

and by virtue of the Lorentz transformation:

$$\delta x' = kl(\delta x + \epsilon \delta t), \quad \delta y' = l\delta y, \quad \delta z' = l\delta z, \quad \delta t' = kl(\delta t + \epsilon x).$$

where, assuming $\delta t = 0$, the relationships:

$$\begin{cases} \delta x' = \delta U' + \xi \delta t' = kl\delta U, \\ \delta y' = \delta V' + \eta' \delta t' = l\delta V, \\ \delta t' = kl\epsilon\delta U. \end{cases}$$

44

Let us note that

$$\xi' = \frac{\xi + \epsilon}{1 + \xi\epsilon}, \quad \eta' = \frac{\eta}{k(1 + \xi\epsilon)};$$

we will obtain, by replacing $\delta t'$ by its value,

$$kl(1 + \xi\epsilon)\delta U = \delta U'(1 + \xi\epsilon) + (\xi + \epsilon)kl\epsilon\delta U,$$
$$l(1 + \xi\epsilon)\delta V = \delta V'(1 + \xi\epsilon) + \eta l\epsilon\delta U.$$

If we remember the definition of k, we will learn from this relationship:

$$\delta U = \frac{k}{l}\delta U' + \frac{k\epsilon}{l}\xi\delta U',$$
$$\delta V = \frac{1}{l}\delta V' + \frac{k\epsilon}{l}\eta\delta U',$$

and similarly

$$\delta W = \frac{1}{l}\delta W' + \frac{k\epsilon}{l}\zeta\delta U',$$

whence

$$\sum X\delta U = \frac{1}{l}(kX\delta U' + Y\delta V' + Z\delta W') + \frac{k\epsilon}{l}\delta U'\sum X\xi. \qquad (3)$$

But, by virtue of equations (2) we must have:

$$\int \sum X'\delta U'dt'd\tau' = \int \sum X\delta U\,dtd\tau = \frac{1}{l^4}\sum X\delta U\,dt'd\tau'.$$

By replacing $\sum X\delta U$ by its value (3) and the identity, we obtain

$$X' = \frac{k}{l^5}X + \frac{k\epsilon}{l^5}\sum X\xi, \quad Y' = \frac{1}{l^5}Y, \quad Z' = \frac{1}{l^5}Z.$$

These are equations (11) of paragraph 1. The principle of least action therefore leads us to the same result as the analysis of paragraph 1. If we refer back to formulas (1), we see that $\sum f^2 - \sum \alpha^2$ is not altered by the Lorentz transformation, except for a constant factor; it is not the case for the expression $\sum f^2 + \sum \alpha^2$, which represents the energy. If we limit ourselves to the case in which ϵ is small enough that we can neglect the square so that $k = 1$ and if we also assume $l = 1$, we find:

$$\sum f'^2 = \sum f^2 + 2\epsilon(g\gamma - h\beta)$$
$$\sum \alpha'^2 = \sum \alpha^2 + 2\epsilon(g\gamma - h\beta)'$$

or, by addition

$$\sum f'^2 + \sum \alpha'^2 = \sum f^2 + \sum \alpha^2 + 4\epsilon(g\gamma - h\beta).$$

4. The Lorentz Group

It is important to note that the Lorentz transformations form a group. Indeed, if we pose:

$$x' = kl(x + \epsilon t), \quad y' = ly, \quad z' = lz, \quad t' = kl(t + \epsilon x),$$

and on the other hand

$$x'' = k'l'(x' + \epsilon't'), \quad y'' = l'y', \quad z'' = l'z', \quad t'' = k'l'(t' + \epsilon'x'),$$

with

$$k^{-2} = 1 - \epsilon^2, \quad k'^{-2} = 1 - \epsilon'^2$$

we will obtain:

$$x'' = k''l''(x + \epsilon''t), \quad y'' = l''y, \quad z'' = l''z, \quad t'' = k''l''(t + \epsilon''x),$$

with

$$\epsilon'' = \frac{\epsilon + \epsilon'}{1 + \epsilon\epsilon'}, \quad l'' = ll', \quad k'' = kk'(1 + \epsilon\epsilon') = \frac{1}{\sqrt{1 - \epsilon''^2}}.$$

If we give l the value 1 and assume that ϵ is infinitely small,

$$x' = x + \delta x, \quad y' = y + \delta y, \quad z' = z + \delta z, \quad t' = t + \delta t,$$

we will obtain:

$$\delta x = \epsilon t, \quad \delta y = \delta z = 0, \quad \delta t = \epsilon x.$$

This is the infinitesimal transformation that generates the group, which I will call the T_1 transformation and which according to Lie notation can be written:

$$t\frac{d\phi}{dx} + x\frac{d\phi}{dt} = T_1.$$

If we assume $\epsilon = 0$ and $l = 1 + \delta l$, we would find instead

$$\delta x = x\delta l, \quad \delta y = y\delta j, \quad \delta z = z\delta l, \quad \delta t = t\delta l$$

and we would have another infinitesimal transformation T_0 of the group (assuming that l and ϵ are considered independent variables) and we would have with the Lie notation:

$$T_0 = x\frac{d\phi}{dx} + y\frac{d\phi}{dy} + z\frac{d\phi}{dz} + t\frac{d\phi}{dt}.$$

But we could make the y-axis or the z-axis play the particular role that we were making the x-axis play; this would give us two other infinitesimal transformations:

$$T_2 = t\frac{d\phi}{dy} + y\frac{d\phi}{dt},$$
$$T_3 = t\frac{d\phi}{dz} + z\frac{d\phi}{dt},$$

that would not alter Lorentz's equations either. We can configure the combinations imagined by Lie, such as

$$[T_1, T_2] = x\frac{d\phi}{dy} - y\frac{d\phi}{dx};$$

but it is easy to see that this transformation is equivalent to a change of coordinate axes, the axes rotating at a very small angle around the z-axis. We should therefore not be surprised if such a change does not alter the shape of the Lorentz equations, which are obviously independent of the choice of axes. We are therefore led to consider a continuous group which we will name the Lorentz group and which will admit as infinitesimal transformations:

1° the transformation T_0 which will be interchangeable with all of the others;

2° the three transformations T_1, T_2, T_3;

3° the three rotations $[T_1, T_2], [T_2, T_3], [T_3, T_1]$.

Any transformation of this group can always be decomposed into a transformation of the form:

$$x' = lx, \quad y' = ly, \quad z' = lz, \quad t' = lt$$

and a linear transformation that does not alter the quadratic form.

$$x^2 + y^2 + z^2 - t^2$$

We can still generate our group in another way. Any transformation of the group can be seen as a transformation of form:

$$x' = kl(x + \epsilon t), \quad y' = ly, \quad z' = lz, \quad t' = kl(t + \epsilon x) \tag{1}$$

preceded and followed by a suitable rotation. But for our purpose, we must consider only a part of the transformations of this group; we must assume that l is a function of ϵ, and the point is to choose this function, so that this part of the group, which I will call P, still forms a group.

Let us rotate the system by 180° around the y-axis, we should find again a transformation that should still belong to P. But this is the same as changing the sign of x, x', z and z'; we then find:

$$x' = kl(x - \epsilon t)' \quad y' = ly, \quad z' = lz, \quad t' = kl(t - \epsilon x) \tag{2}$$

So l doesn't change when you change ϵ to $-\epsilon$. On the other hand, if P is a group, the inverse substitution of (1) which is written:

$$x' = \frac{k}{l}(x - \epsilon t), \quad y' = \frac{y}{l}, \quad z' = \frac{z}{l}, \quad t' = \frac{k}{l}(t - \epsilon x), \tag{3}$$

must also belong to P; it must therefore be identical to (2), meaning that

$$l = \frac{1}{l}.$$

So we will have $l = 1$.

5. Langevin Waves

Mr. Langevin has put into a particularly elegant form the formulas that define the electromagnetic field produced by the motion of a single electron. Let us go back to equations

$$\Box \psi = -\rho, \quad \Box F = -\rho \xi. \tag{1}$$

We know that we can integrate them through the delayed potentials and that we have:

$$\psi = \frac{1}{4\pi} \int \frac{\rho_1 d\tau_1}{r}, \quad F = \frac{1}{4\pi} \int \frac{\rho_1 \xi_1 d\tau_1}{r}. \tag{2}$$

In these formulas, we have:

$$dz_1 = dx_1 dy_1 dz_1, \quad r^2 = (x - x_1)^2 + (y - y_1)^2 + (z - z_1)^2,$$

$$x_1 = x_0 + U, \quad y_1 = y_0 + V, \quad z_1 = z_0 + W$$

while ρ_1 and ξ_1 are the values of ρ and ξ at point x_1, y_1, z_1, and instant

$$t_1 = t - r.$$

Let: x_0, y_0, z_0 be the coordinates of an electron occurrence at instant t; and

$$x_1 = x_0 + U, \quad y_1 = y_0 + V, \quad z_1 = z_0 + W$$

its coordinates at time t_1. U, V, W are functions of x_0, y_0, z_0, so we can write:

$$dx_1 = dx_0 + \frac{dU}{dx_0} dx_0 + \frac{dU}{dy_0} dy_0 + \frac{dU}{dz_0} dz_0 + \xi_1 dt_1$$

and assuming t constant, as well as x, y and z:

$$dt_1 = dt_1 = + \sum \frac{x - x_1}{r} dx_1.$$

So we can write:

$$dx_1 \left(1 + \xi_1 \frac{x_1 - x}{r}\right) + dy_1 \xi_1 \frac{y_1 - y}{r} + dz_1 \xi_1 \frac{z_1 - z}{r} = dx_0 \left(1 + \frac{dU}{dx_0}\right) + dy_0 \frac{dU}{dy_0} + dz_0 \frac{dU}{dz_0}$$

with the two other equations that can be deduced by circular permutation. So we have:

$$d\tau_1 \left| 1 + \xi_1 \frac{x_1 - x}{r}, \xi_1 \frac{y_1 - y}{r}, \xi_1 \frac{z_1 - z}{r} \right| = d\tau_0 \left| 1 + \frac{dU}{dx_0}, \frac{dU}{dy_0}, \frac{dU}{dz_0} \right|, \tag{3}$$

posing

$$d\tau_0 = dx_0 \, dy_0 \, dz_0.$$

Let's study the determinants that appear in both members of (3) and first in the 1st member; if we try to develop it, we see that the terms of the second and third degree in relation to ξ_1, η_1, ζ_1 disappear and that the determinant is equal to

$$1 + \xi_1 \frac{x_1 - x}{r} + \eta_1 \frac{y_1 - y}{r} + \zeta_1 \frac{z_1 - z}{r} = 1 + \omega,$$

ω designating the radial component of velocity ξ_1, η_1, ζ_1, i.e. the component directed along the radius that goes from point x, y, z to point x_1, y_1, z_1.

To obtain the second determinant, I consider the coordinates of the different occurrences of the electron at an instant t', which is the same for all occurrences, but in such a way that for the occurrence that I consider we have $t_1 = t_1'$. The coordinates of an occurrence will then be:

$$x_1' = x_0 + U', \quad y_1' = y_0 + V', \quad z_1' = z_0 + W',$$

U', V', W' being what U, V, W become when we replace t_1 by t_1' in it; since t_1' is the same for all occurrences, we will have:

$$dx_1' = dx_0 \left(1 + \frac{dU'}{dx_0}\right) + dy_0 \frac{dU'}{dy_0} + dz_0 \frac{dU'}{dz_0}$$

and consequently

$$d\tau_1' = d\tau_0 |1 + \frac{dU'}{dx_0}, \frac{dU'}{dy_0}, \frac{dU'}{dz_0}|,$$

posing

$$d\tau_1' = dx_1' dy_1' dz_1'.$$

But the element of electrical charge is

$$d\mu_1 = \rho_1 d\tau_1'$$

and moreover, *for the molecule considered*, we have $t_1 = t_1'$ and therefore $\frac{dU'}{dx_0} = \frac{dU}{dx_0}$ etc.; so we can write:

$$d\mu_1 = \rho_1 d\tau_0 |1 + \frac{dU}{dx_0}, \frac{dU}{dy_0}, \frac{dU}{dz_0}|$$

so that equation (3) becomes:

$$\rho_1 d\tau_1 (1 + \omega) = d\mu_1$$

and equations (2):

$$\psi = \frac{1}{4\pi} \int \frac{d\mu_1}{r(1+\omega)}, \quad F = \frac{1}{4\pi} \int \frac{\xi_1 d\mu_1}{r(1+\omega)}.$$

If we are dealing with a single electron, our integrals will be reduced to a single element, as long as we consider only points x, y, z far enough apart so that r and ω have approximately the same value for all the points of the electron. The potentials ψ, F, G, H will depend on the position of this electron, and also on its velocity, because not only ξ_1, η_1, ζ_1 appear in the numerator in F, G, H, but the radial component ω appears in the denominator. This is of course its position and velocity at time t_1.

The partial derivatives of ϕ, F, G, H with respect to t, x, y, z (and therefore the electric and magnetic fields) will also depend on its acceleration. Moreover, they will depend *linearly* on it, since in these derivatives this acceleration is introduced as a result of a single differentiation.

Langevin was thus led to distinguish in the electric and magnetic fields the terms that do not depend on the acceleration (this is what he named the velocity wave) and those that are proportional to the acceleration (this is what he named the acceleration wave).

Calculation of these two waves is made easier by the Lorentz transformation. We can indeed apply this transformation to the system, so that the velocity of the single electron considered becomes zero. We will take for the x-axis the direction of this velocity before the transformation, so that at instant t_1,

$$\eta_1 = \zeta_1 = 0$$

and we will take $\epsilon = -\xi$, so that

$$\xi_1' = \eta_1' = \zeta_1' = 0.$$

We can therefore bring the calculation of the two waves back to the case where the speed of the electron is zero. Let's start with the velocity wave; we first notice that this wave is the same as if the motion of the electron was uniform. If the electron velocity is zero, we have:

$$\omega = 0, \quad F = G = H = 0, \quad \psi = \frac{\mu_1}{4\pi r},$$

μ_1 being the electric charge of the electron. The velocity having been reduced to zero by the Lorentz transformation, we then have:

$$F' = G' = H' = 0, \quad \psi' = \frac{\mu_1}{4\pi r'},$$

r' being the distance from point x', y', z' to point x_1', y_1', z_1', and consequently:

$$\alpha' = \beta' = \gamma' = 0,$$

$$f' = \frac{\mu'(x' - x_1')}{4\pi r'^3}, \quad g' = \frac{\mu'(y' - y_1')}{4\pi r'^3}, \quad h' = \frac{\mu'(z' - z_1')}{4\pi r'^3}.$$

Let us now do the transformation inverse to that of Lorentz to find the true field corresponding to a velocity $\epsilon, 0, 0$. We find, by referring to equations (9) and (3) of paragraph 1:

$$\begin{cases} \alpha = 0, \quad \beta = \epsilon h, \quad \gamma = -\epsilon g, \\ f = \frac{\mu_1 k l^3}{4\pi r'^3}(x + \epsilon t - x_1 - \epsilon t_1), \quad g = \frac{\mu_1 k l^3}{4\pi r'^3}(y - y_1), \quad h = \frac{\mu_1 k l^3}{4\pi r'^3}(z - z_1). \end{cases} \quad (4)$$

We can see that the magnetic field is perpendicular to the x-axis (direction of velocity) and to the electric field, and that the electric field is pointing towards point:

$$x_1 + \epsilon(t_1 - t), \ y_1, \ z_1. \quad (5)$$

If the electron continued to move in a rectilinear and uniform motion with the velocity it had at instant t_1, i.e. with velocity $-\epsilon, 0, 0$ this point (5) would be the one it would

50

occupy at instant t. Let's move on to the acceleration wave; we can, thanks to the Lorentz transformation, bring its determination back to the case when the velocity is null. This is the cas which is realized if we imagine an electron performing oscillations of very small amplitude, at very small velocities, but that the accelerations are finite. This brings us back to the field that was studied in the famous Memoir by Hertz entitled *Die Kräfte elektrischer Schwingungen nach der Maxwell'schen Theorie* and this, for a point far away. Under these conditions:

1^0 The two electric and magnetic fields are mutually equal.
2^0 They are perpendicular to each other.
3^0 They are perpendicular to the normal to the wave sphere, i.e. to the sphere whose centre is point x_1, y_1, z_1.

I say that these three properties will still exist when the velocity is not zero, and for that, I only need to show that they are not altered by the Lorentz transformation. Indeed, let A be the intensity common to bothy fields and let

$$(x - x_1) = r\lambda, \quad (y - y_1) = r\mu, \quad (z - z_1) = r\nu, \quad \lambda^2 + \mu^2 + \nu^2 = 1,$$

These properties will be expressed in terms of equalities:

$$A^2 = \sum f^2 = \sum \alpha^2, \quad \sum f\alpha = 0, \quad \sum f(x - x_1) = 0, \quad \sum \alpha(x - x_1) = 0,$$

$$\sum f\lambda = 0, \sum \alpha\lambda = 0;$$

which also means that

$$\frac{b}{A}, \quad \frac{g}{A}, \quad \frac{h}{A};$$
$$\frac{\alpha}{A}, \quad \frac{\beta}{A}, \quad \frac{\gamma}{A};$$
$$\lambda, \quad \mu, \quad \nu,$$

are the directing cosines of three rectangular directions, and we deduce their relationships:

$$f = \beta\nu - \gamma\mu, \quad \alpha = h\mu - g\nu,$$

or

$$fr = \beta(z - z_1) - \gamma(y - y_1), \quad \alpha r = h(y - y_1) - g(z - z_1), \tag{6}$$

with the equations that can be deduced from them by symmetry. If we go back to equations (3) of paragraph 1, we find:

$$\begin{cases} x' - x_1' = kl[(x - x_1) + \epsilon(t - t_1)] = kl[(x - x_1) + \epsilon r], \\ y' - y_1' = l(y - y_1), \\ z' - z_1' = l(z - z_1). \end{cases} \tag{7}$$

We found earlier in paragraph 3:

$$l^4 \left(\sum f'^2 - \sum \alpha'^2 \right) = \sum f^2 - \sum \alpha^2.$$

So $\sum f^2 = \sum \alpha^2$ leads to $\sum f'^2 = \sum \alpha'^2$. On the other hand, starting from equations (9) of paragraph 1, we find:

$$l^4 \sum f'\alpha' = \sum f\alpha,$$

This shows that $\sum f\alpha = 0$ leads to $\sum f'\alpha' = 0$. I say now that

$$\sum f'(x' - x_1') = 0, \quad \sum \alpha'(x' - x_1') = 0. \tag{8}$$

Indeed, by virtue of equations (7) [as well as from Equations 9 of paragraph 1] the first members of the two equations (8) can be written respectively:

$$\frac{k}{l} \sum f(x - x_1) + \frac{k\epsilon}{l}[fr + \gamma(y - y_1) - \beta(z - z_1)],$$
$$\frac{k}{l} \sum \alpha(x - x_1) + \frac{k\epsilon}{l}[\alpha r + h(y - y_1) - g(z - z_1)].$$

So they cancel each other out according to the equations

$$\sum f(x - x_1) = \sum \alpha(x - x_1) = 0$$

and according to equations (6). Yet that is precisely what was meant to be demonstrated.

One can arrive at the same result by simple considerations of homogeneity. Indeed, ψ, F, G, H are functions of $(x - x_1), (y - y_1), (z - z_1), \xi_1 = \frac{dx_1}{dt_1}, \eta_1 = \frac{dy_1}{dt_1}, \zeta_1 = \frac{dz_1}{dt_1}$, being homogeneous in degree -1 with respect to $x, y, z, t, x_1, y_1, z_1, t_1$ and to their differentials.

So the derivatives of ψ, F, G, H with respect to x, y, z, t (and consequently also both fields $f, g, h; \alpha, \beta, \gamma$), will be homogeneous by degree -2 with respect to the same quantities, if we remember that the relationship

$$t - t_1 = r = \sqrt{\sum})x - x_1)^2$$

is homogeneous with respect to these quantities.

But these derivatives or fields depend on the $(x - x_1)$, the velocities $\frac{dx_1}{dt_1}$, and the accelerations $\frac{d^2x_1}{dt_1^2}$; they consist of a term independent of accelerations (velocity wave) and a term linear with respect to they accelerations (acceleration waves). But $\frac{dx_1}{dt_1}$ is homogeneous of degree 0 and $\frac{d^2x_1}{dt_1^2}$ of degree -1; hence it follows that the velocity wave is homogeneous of degree -2 with respect to $x - x_1, y - y_1, z - z_1$, and the acceleration wave is homogeneous of degree -1. Therefore, at a very distant point the acceleration wave is preponderant and can therefore be regarded as merging with the total wave. Furthermore, the law of homogeneity shows us that the acceleration wave is similar to itself at a distant point and at any point. It is therefore at any point similar to the total wave at a distant point. However, at a distant point the disturbance can only propagate by plane waves, so the two fields must be equal, perpendicular to each other and perpendicular to the direction of propagation.

I will simply refer for more details to Mr Langevin's dissertation in the *Journal de Physique* (Year 1905).

6. Contraction of Electrons

Let us suppose a single electron animated by a rectilinear and uniform translational motion. According to what we have just seen, we can, thanks to the Lorentz transformation, bring back the study of the field determined by this electron to the case where the electron is immobile; the Lorentz transformation thus replaces the real electron in motion by an immobile ideal electron.

Let α, g, h be the real field; and let α', f', g', h' be what the field becomes after the Lorentz transformation, so that the ideal field α', f', corresponds to the case of a stationary electron; we have:

$$\alpha' = \beta' = \gamma' = 0, \quad f' = \frac{d\psi'}{dx'}, \qquad g' = \frac{d\psi'}{dy'}, \quad h' = -\frac{d\psi'}{dz'}$$

and for the actual field [in accordance with formulae (9) of paragraph 1]:

$$\begin{cases} \alpha = 0, \quad \beta = \epsilon h, \quad \gamma = -\epsilon g \\ f = l^2 f', \quad g = kl^2 g', \quad h = kl^2 h'. \end{cases} \tag{1}$$

The next step is to determine the total energy due to the motion of the electron, the corresponding action and the amount of electromagnetic momentum, in order to be able to calculate the electromagnetic masses of the electron. For a distant point, it is enough to consider the electron as reduced to a single point; one is thus brought back to the formulas (4) of the previous paragraph which generally may be suitable. But here they are not sufficient, because the energy is mainly located in the parts of the ether closest to the electron.

Several hypotheses can be made from this perspective.

According to that of Abraham, electrons are spherical and non-deformable.

Then, when applying the Lorentz transformation, since the real electron would be spherical, the ideal electron would become an ellipsoid. The equation of this ellipsoid would be according to paragraph 1:

$$k^2(x' - \epsilon t' - \xi t' + \epsilon \xi x')^2 + (y' - \eta k t' - \xi t' + \eta k \epsilon x')^2 + (z' - \zeta k t' + \zeta k \epsilon x')^2 = l^2 r^2$$

But here, we have:

$$\xi + \epsilon = \eta = \zeta = 0, \quad 1 + \epsilon \xi = 1 - \epsilon^2 = \frac{1}{k^2},$$

so that the ellipsoid equation becomes:

$$\frac{x'^2}{k^2} + y'^2 + z'^2 = l^2 r^2.$$

If the radius of the real electron is r, then the axes of the ideal electron would be:

$$klr, \quad lr, \quad lr$$

In Lorentz's hypothesis, contrariwise, moving electrons would be deformed in such a way that it would be the real electron that would become an ellipsoid, while the ideal immobile electron would always be a sphere of radius r; the axes of the real electron would then be:

$$\frac{r}{lk}, \quad \frac{r}{l}, \quad \frac{r}{l}.$$

Let us designate by

$$A = \frac{1}{2} \int f^2 d\tau,$$

the longitudinal electrical energy; by

$$B = \frac{1}{2} \int (g^2 + h^2) d\tau$$

the transverse electrical energy; by

$$C = \frac{1}{2} \int (\beta^2 + \gamma^2) d\tau$$

the transverse magnetic energy. There is no longitudinal magnetic energy, since $\alpha = \alpha' = 0$. Let's designate by A', B', C' the corresponding quantities in the ideal system. We find first:

$$C' = 0, C = \epsilon^2 B.$$

On the other hand, we can observe that the real field depends only on $x + \epsilon t, y$ and z, and write:

$$d\tau = d(x + \epsilon t) dy dz,$$
$$d\tau' = dx' dy' dz' = kl^3 d\tau;$$

whence

$$A' = \frac{k}{l} A, \quad B' = \frac{1}{kl} B, \quad A = \frac{lA'}{k}, \quad B = klB'.$$

Under the Lorentz hypothesis we have $B' = 2A'$, and A', inversely proportional to the electron radius, is a constant independent of the velocity of the real electron; thus we find for the total energy:

$$A + B + C = A'lk(3 + \epsilon^2)$$

and for action (per unit of time):

$$A + B - C = \frac{3A'l}{k}$$

Now let's calculate the electromagnetic momentum; we'll find

$$D = \int (g\gamma - h\beta) d\tau = -\epsilon \int (g^2 + h^2) d\tau = -2\epsilon B = -4\epsilon klA'.$$

But there must be some relationship between the energy $E = A + B + C$, the action per unit of time $H = A + B - C$, and momentum D. The first of these relationships is:

$$E = H = -\epsilon \frac{dH}{d\epsilon},$$

the second is

$$\frac{dD}{d\epsilon} = -\frac{1}{\epsilon}\frac{dE}{d\epsilon};$$

whence:

$$D = \frac{dH}{d\epsilon}, \quad E = H - \epsilon D. \tag{2}$$

The second of equations (2) is always satisfied; but the first is only satisfied if

$$l = (1 - \epsilon^2)^{\frac{1}{6}} = k^{-\frac{1}{3}},$$

i.e. if the volume of the ideal electron is equal to that of the real electron, or if the volume of the electron is constant; this is Langevin's hypothesis.

This is in contradiction with the result of paragraph 4 and with the result obtained by Lorentz by another means. It is this contradiction that needs to be explained.

Before addressing this explanation, I observe that, whatever the hypothesis adopted we will have

$$H = A + B - C = \frac{l}{k}(A' + B'),$$

Or, since $C' = 0$;

$$H = \frac{l}{k}H', \tag{3}$$

We can relate this result to the equation J = J' obtained in paragraph 3. Indeed we have:

$$J = \int H dt, \quad J' \int H' dt'.$$

We will observe that the state of the system depends only on $x + \epsilon t, y$ and z, i.e. on x', y', z', and that we have:

$$t' = \frac{l}{k}t + \epsilon x'$$
$$dt' = \frac{l}{k}dt. \tag{4}$$

Comparing equations (3) and (4) we find $J = J'$. Let us place ourselves in an unspecified hypothesis, which could be, either that of Lorentz or that of Abraham, or that of Langevin, i.e. an intermediate hypothesis. Let us pose:

$$r, \theta r, \theta r$$

as the three axes of the real electron; those of the ideal electron will be:

$$klr, \theta lr, \theta lr.$$

then $A' + B'$ will be the electrostatic energy due to an ellipsoid having for axes $klr, \theta lr, \theta lr$.

Whether we assume the electricity spread over the surface of the electron as known to a conductor, or uniformly distributed within this electron; this energy will be of form:

$$A' + B' = \frac{\varphi\left(\frac{\theta}{l}\right)}{klr},$$

where φ is a known function.

The Abraham hypothesis is to assume

$$r = const., \quad \theta = 1.$$

That of Lorentz:

$$l = 1, \quad kr = const., \quad \theta = k.$$

That of Langevin:

$$l = k^{-\frac{1}{3}}, \quad k = \theta, \quad krl = const.$$

We then find:

$$H = \frac{\varphi\left(\frac{\theta}{k}\right)}{k^2 r}.$$

Abraham finds, with the exception of the notations (*Göttinger Nachrichten*, 1902, p. 37):

$$H = \frac{a}{r} \frac{1 - \epsilon^2}{\epsilon} \log \frac{1 + \epsilon}{1 - \epsilon},$$

a being a constant. However, in the Abraham hypothesis, we have $\theta = 1$; therefore:

$$\varphi\left(\frac{1}{k}\right) = ak^2 \frac{1 - \epsilon^2}{\epsilon} \log \frac{1 + \epsilon}{1 - \epsilon} = \frac{a}{\epsilon} \log \frac{1 + \epsilon}{1 - \epsilon}, \tag{5}$$

which defines the function φ.

Having said that, let us imagine that the electron is subjected to a binding, so that there is a relation between r and θ; in the hypothesis of Lorentz this relation would be $\theta r = const.$, in that of Langevin $\theta^2 r^3 = const.$. We will suppose in a more general way

$$r = b\theta^m$$

b being a constant; whence:

$$H = \frac{1}{bk^2} \theta^{-m} \varphi\left(\frac{\theta}{k}\right).$$

What form will the electron take when the velocity becomes $-\epsilon t$, *if we do not assume the intervention of forces other than those of bonding?* This form will be defined by equality:

$$\frac{\partial H}{\partial \theta} = 0, \tag{6}$$

or

$$-m\theta^{-m-1}\varphi + \theta^{-m}k^{-1}\varphi' = 0,$$

or

$$\frac{\varphi'}{\varphi} = \frac{mk}{\theta}.$$

If we want equilibrium to occur so that $\theta = h$, it is necessary that $\frac{\theta}{k} = 1$ and the negative logrithmic derivative of φ is equal to m. If we develop $\frac{1}{k}$ and the right hand side of (5) in powers of ϵ, equation (5) becomes:

$$\varphi\left(1 - \frac{\epsilon^2}{2}\right) = \alpha\left(1 + \frac{\epsilon^2}{3}\right),$$

neglecting the higher powers of ϵ. Differentiating, we obtain:

$$-\epsilon\varphi'\left(1 - \frac{\epsilon^2}{2}\right) = \frac{2}{3}\epsilon a.$$

For $\epsilon = 0$, i.e. when the argument of φ is equal to 1, these equations become:

$$\varphi = a, \quad \varphi' = -\frac{2}{3}a, \quad \frac{\varphi'}{\varphi} = -\frac{2}{3}. \tag{7}$$

We must therefore have $m = -\frac{2}{3}$ in conformity with the hypothesis of Langevin. This result must be compared with the result for the first equation (2), from which it does not differ in reality. Indeed, if we suppose that any element $d\tau$ of the electron is subjected to a force $Xd\tau$ parallel to the x-axis, X being the same for all elements, we will then have, according to the definition of momentum:

$$\frac{dD}{dt} = \int Xd\tau.$$

On the other hand, the principle of least action gives us:

$$\delta J = \int X\delta U d\tau dt, \quad J = \int Hdt, \quad \delta J = \int D\delta U dt,$$

where δU is the displacement of the electron's centre of gravity; H depends on θ and ϵ, if we assume that r is linked to θ by the binding equation, then we have:

$$\delta J = \int \left(\frac{\partial H}{\partial \epsilon}\delta\epsilon + \frac{\partial H}{\partial \theta}\delta\theta\right)dt.$$

On the other hand, $\delta\epsilon = -\frac{d\delta U}{dt}$; where by integrating by parts

$$\int D\delta\epsilon dt = \int D\delta U dt,$$

or

$$\int \left(\frac{\partial H}{\partial \epsilon}\delta\epsilon + \frac{\partial H}{\partial \theta}\delta\theta\right)dt = \int D\delta\epsilon dt,$$

whence

$$D = \frac{\partial H}{\partial \epsilon}, \quad \frac{\partial H}{\partial \theta} = 0.$$

But the derivative $\frac{dH}{d\epsilon}$, containing in the right hand side of equation (2), is the derivative by supposing θ as a function of ϵ, so that

$$\frac{dH}{d\epsilon} = \frac{\partial H}{\partial \epsilon} + \frac{\partial H}{\partial \theta}\frac{d\theta}{d\epsilon}$$

Equation (2) is therefore equivalent to equation (6).

The conclusion is that if the electron is subjected to a bond between its three axes, *and if no other force intervenes apart from the bonding forces*, the shape that this electron will take, when it is animated with a uniform velocity, cannot be such that the corresponding ideal electron is a sphere, unless the bond is that the volume is constant, in accordance with the Langevin hypothesis.

This raises the following question: what additional forces, other than the binding forces, would be needed to account for the Lorentz law or, more generally, for any law other than the Langevin law?

The simplest hypothesis, and the first one we had to examine, is that these additional forces derive from a special potential deriving from the three axes of the ellipsoid, and therefore from θ and r; i.e. $F(\theta, r)$ this potential; in this case the action, will be expressed:

$$J = \int [H + F(\theta, r)]dt$$

and the equilibrium conditions will be written:

$$\frac{dH}{d\theta} + \frac{dF}{d\theta} = 0, \quad \frac{dH}{dr} + \frac{dF}{dr} = 0. \tag{8}$$

If we assume r and θ linked by the relation $r = b\theta^m$, we can look at r as a function of θ, consider F as depending only on θ and keep only the first equation (8) with:

$$H = \frac{\varphi}{bk^2\theta^m}, \quad \frac{dH}{d\theta} = -\frac{m\varphi}{bk^2\theta^{m+1}} + \frac{\varphi'}{bk^3\theta^m}.$$

It is necessary that, for $k = \theta$, equation (8) be satisfied; this gives, taking into account equations (7):

$$\frac{dF}{d\theta} = \frac{ma}{b\theta^{m+3}} + \frac{2}{3}\frac{a}{b\theta^{m+3}},$$

whence:

$$F = -\frac{a}{b\theta^{m+2}}\frac{m+\frac{2}{3}}{m+2}$$

and under the Lorentz hypothesis, in which $m = -1$:

$$F = \frac{a}{3b\theta}.$$

Let us now assume that there is no bond and, considering r and θ as two independent variables, let us keep both equations (8); then we will have:

$$H = \frac{\varphi}{k^2 r}, \quad \frac{dH}{d\theta} = \frac{\varphi'}{k^3 r}, \quad \frac{dH}{dr} = -\frac{\varphi}{k^2 r^2}.$$

Equations (8) must be satisfied for $k = \theta, r = b\theta^m$; which gives:

$$\frac{dF}{dr} = \frac{a}{b^2\theta^{2m+2}}, \quad \frac{DF}{d\theta} = \frac{2}{3}\frac{a}{b\theta^{m+3}}. \tag{9}$$

One way to meet these conditions is to pose:

$$F = Ar^\alpha\theta^\beta, \tag{10}$$

A, α, β being constants; equations (9) must be satisfied for $k = \theta, r = b\theta^m$, which gives :

$$Aab^{\alpha-1}\theta^{m\alpha-m+\beta} = \frac{a}{b^2\theta^{2m+2}}, \quad A\beta b^\alpha\theta^{m\alpha+\beta-1} = \frac{2}{3}\frac{a}{b\theta^{m+3}}.$$

By identifying we find

$$\alpha = 3\gamma, \quad \beta = 2\gamma, \quad \gamma = -\frac{m+2}{3m+2}, \quad A = \frac{a}{\alpha b^{\alpha-1}}. \tag{11}$$

But the volume of the ellipsoid is proportional to $r^3\theta^2$, so the extra potential is proportional to the power γ of the electron volume.

In the Lorentz hypothesis, we have $m = -1, \gamma = 1$.

We therefore recover the Lorentz hypothesis on the condition of adding an additional potential proportional to the volume of the electron.

The Langevin hypothesis corresponds to $\gamma = \infty$.

7. Quasi-stationary motion

It remains to be seen whether this hypothesis on the contraction of electrons accounts for the impossibility of demonstrating absolute motion, and I will begin by studying the quasi-stationary motion of an isolated electron, or subject only to the action of other distant electrons.

We know that quasi-stationary motion is a motion in which the variations in velocity are slow enough that the magnetic and electrical energies due to the motion of the electron differ little from what they would be in uniform motion; we also know that it is from this notion of quasi-stationary motion that Abraham arrived at that of transverse and longitudinal electromagnetic masses.

I think I should clarify. Let H be our action per unit of time:

$$H = \frac{1}{3}\int\left(\sum f^2 - \sum \alpha^2\right)d\tau,$$

where for the moment we only consider electric and magnetic fields due to the motion of a single electron. In the previous paragraph, considering motion as uniform, we looked at H as dependent on the velocity ξ, η, ζ of the center of gravity of the electron (these three components, in the previous paragraph, had the values $-\epsilon, 0, 0$) and the parameters r and θ that define the shape of the electron.

But if the motion is no longer uniform, H will depend not only on the values of $\xi, \eta, \zeta, r, \theta$ at the moment in question, but on the values of these same quantities at other instants which may differ from them by quantities of the same order as the time taken by light to go from one point to another of the electron; in other words, H will depend not only on $\xi, \eta, \zeta, r, \theta$, but on their derivatives of all orders with respect to time.

Well, the motion will be said to be quasi-stationary when the partial derivatives of H with respect to the successive derivatives of $\xi, \eta, \zeta, r, \theta$ will be negligible in front of the partial derivatives of H with respect to the $\xi, \eta, \zeta, r, \theta$ quantities themselves.

The equations of such a motion can be written:

$$
\begin{cases}
\dfrac{dH}{d\theta} + \dfrac{dF}{d\theta} = \dfrac{dH}{dr} + \dfrac{dF}{dr} = 0, \\[2mm]
\dfrac{d}{dt}\dfrac{dH}{d\xi} = -\int X d\tau, \quad \dfrac{d}{dt}\dfrac{dH}{d\eta} = -\int Y d\tau, \quad \dfrac{d}{dt}\dfrac{dH}{d\zeta} = -\int Z d\tau.
\end{cases}
\tag{1}
$$

In these equations, F has the same meaning as in the previous paragraph; X, Y, Z are the components of the force that acts on the electron: this force being due solely to the electric and magnetic fields produced by the other electrons. Observe that H depends on ξ, η, ζ only through the combination

$$
V = \sqrt{\xi^2 + \eta^2 + \zeta^2},
$$

that is, of the magnitude of the velocity; thus, by naming again D the momentum:

$$
\frac{dH}{d\xi} = \frac{dH}{dV}\frac{\xi}{V} = -D\frac{\xi}{V},
$$

whence:

$$
-\frac{d}{dt}\frac{dH}{d\xi} = \frac{D}{V}\frac{d\xi}{dt} - D\frac{\xi}{V^2}\frac{dV}{dt} + \frac{dD}{dV}\frac{\xi}{V}\frac{dV}{dt},
\tag{2}
$$

$$
-\frac{d}{dt}\frac{dH}{d\eta} = \frac{D}{V}\frac{d\eta}{dt} - D\frac{\eta}{V^2}\frac{dV}{dt} + \frac{dD}{dV}\frac{\eta}{V}\frac{dV}{dt},
\tag{2 bis}
$$

with

$$
V\frac{DV}{dt} = \sum \xi\frac{d\xi}{dt}.
\tag{3}
$$

If we take the current direction of velocity for x-axis, we obtain:

$$
\xi = V, \quad \eta = \zeta = 0, \quad \frac{d\xi}{dt} = \frac{dV}{dt};
$$

equations (2) and (2 bis) become:

$$
-\frac{d}{dt}\frac{DH}{d\xi} - \frac{dD}{dV}\frac{d\xi}{dt}, \quad -\frac{d}{dt}\frac{dH}{d\eta} = \frac{D}{V}\frac{d\eta}{dt}
$$

and the last three equations (1):

$$
\frac{dD}{dV}\frac{d\xi}{dt} = \int X d\tau, \quad \frac{D}{V}\frac{d\eta}{dt} = \int Y d\tau, \quad \frac{D}{V}\frac{d\zeta}{dt} = \int Z d\tau.
\tag{4}
$$

This is why Abraham gave $\frac{dD}{dV}$ the name longitudinal mass and to $\frac{D}{V}$ the name of transverse mass; let us recall that $D = \frac{dH}{dV}$. In the Lorentz hypothesis, we have:

$$
D = -\frac{dH}{dV} = -\frac{\partial H}{\partial V},
$$

$\frac{\partial H}{\partial V}$ representing the derivative with respect to V, after r and θ have been replaced by their values as functions of V from the first two equations (1), and after this substitution we will have

$$H = +A\sqrt{1 - V^2}.$$

We will choose units so that the constant factor A is equal to 1, and I pose $\sqrt{1 - V^2} = h$, hence:

$$H = +h, \quad D = \frac{V}{h}, \quad \frac{dH}{dV} = h^{-3}, \quad \frac{dD}{dV}\frac{1}{V^2} - \frac{D}{V^3} = h^{-3}$$

We will pose again:

$$M = V\frac{dV}{dt} = \sum \xi \frac{d\xi}{dt}, \quad X_1 = \int X d\tau$$

and we will find for the equation of quasi-stationary motion:

$$h^{-1}\frac{d\xi}{dt} + h^{-3}\xi M = X_1. \tag{5}$$

Let's see what happens to these equations by means of the Lorentz transformation. We will pose: $1 + \xi\epsilon = \mu$, and we will have first:

$$\mu\xi' = \xi + \epsilon, \quad \mu\eta' = \frac{\eta}{k}, \quad \mu\zeta' = \frac{\zeta}{k},$$

from which we easily draw

$$\mu h' = \frac{h}{k}.$$

We also have

$$dt' = k\mu dt,$$

whence

$$\frac{\xi'}{dt'} = \frac{d\xi}{dt}\frac{1}{k^3\mu^3}, \quad \frac{\eta'}{dt'} = \frac{d\eta}{dt}\frac{1}{k^2\mu^2} - \frac{d\xi}{dt}\frac{\eta\epsilon}{k^2\mu^3}, \quad \frac{\zeta'}{dt'} = \frac{d\zeta}{dt}\frac{1}{k^2\mu^2} - \frac{d\xi}{dt}\frac{\zeta\epsilon}{k^2\mu^3},$$

from which also:

$$M' = \frac{d\xi}{dt}\frac{\epsilon}{h^2}k^3\mu^4 + \frac{M}{k^3\mu^3}$$

and

$$h'^{-1}\frac{d\xi'}{dt'} + h'^{-3}\xi'M' = \left[h^{-1}\frac{d\xi}{dt} + h^{-3}(\xi + \epsilon)M\right]\mu^{-1}, \tag{6}$$

$$h'^{-1}\frac{d\xi'}{dt'} + h'^{-3}\xi'M' = \left[h^{-1}\frac{d\xi}{dt} + h^{-3}(\xi + \epsilon)M\right]\mu^{-1}. \tag{7}$$

Let us now refer to equations (11 bis) in paragraph 1; we can consider X_1, Y_1, Z_1 as having the same meaning as in equations (5). On the other hand, we have $l = 1$ and $\frac{\rho'}{\rho} = k\mu$; so the equations become

$$\begin{cases} X_1' = \mu^{-1}(X_1 + \epsilon \sum X_1 \xi)' \\ Y_1' = k^{-1}\mu^{-1}Y_1. \end{cases} \tag{8}$$

Let's calculate $\sum X_1 \xi$ Using equations (5), we will find:

$$\sum X_1 \xi = h^{-3}M,$$

whence

$$\begin{cases} X_1' = \mu^{-1}(X_1 + \epsilon h^{-3}M), \\ Y_1' = k^{-1}\mu^{-1}Y_1. \end{cases} \tag{9}$$

By comparing equations (5), (6), (7) and (9), we finally find:

$$\begin{cases} h'^{-1}\dfrac{d\xi'}{dt'} + h'^{-3}\xi'M' = X_1', \\ h'^{-1}\dfrac{d\eta'}{dt'} + h'^{-3}\eta'M' = Y_1'. \end{cases} \tag{10}$$

which shows that the equations of quasi-stationary motion are not altered by the Lorentz transformation; but it still does not prove that the Lorentz hypothesis is the only one that leads to this result.

In order to establish this point, we are going to restrict ourselves, as Lorentz did, to certain specific cases, which will obviously suffice for us to demonstrate a negative proposal.

How do we first extend the assumptions on which the previous calculation was based?

1^0. Instead of assuming $l = 1$ in the transformation of Lorentz, we will assume any l.
2^0. Instead of assuming that F is proportional to volume, and therefore that H is proportional to h, we will assume that F is any function of θ and r, such that [after replacing θ and r by their values as functions of V, taken from the first two equations (1)] H is any function of V.

I first observe that, if we assume $H = h$, we should have $l = 1$; and indeed equations (6) and (7) will remain, except that the right hand sides will be multiplied by $\frac{1}{l}$; so do equations (9), except that the second members will be multiplied by $\frac{1}{l^2}$ and finally equations (10), except that the second term will be multiplied by $\frac{1}{l}$. If the equations of motion are not to be altered by thye Lorentz transformations, i. e., that equations (10) differ from equations (5) only by the accentuation of the letters, then the following must be assumed:

$$l = 1$$

Let us now assume that we have $\eta = \zeta = 0$, where $\xi = V$, $\frac{d\xi}{dt} = \frac{dV}{dt}$; equations (5) will take the form:

$$-\frac{d}{dt}\frac{dH}{d\xi} = \frac{dD}{dV}\frac{d\xi}{dt} = X_1, \qquad -\frac{d}{dt}\frac{dH}{d\eta} = \frac{dD}{dV}\frac{d\eta}{dt} = Y_1. \qquad (5 \text{ bis})$$

We can actually pose:

$$\frac{dD}{dV} = f(V) = f(\xi), \qquad \frac{D}{V} = \varphi(V) = \varphi(\xi).$$

If the equations of motion are not altered by the Lorentz transformation, we should have:

$$f(\xi)\frac{d\xi}{dt} = X_1'$$

$$\varphi(\xi)\frac{d\eta}{dt} = Y_1,$$

$$f(\xi')\frac{d\xi'}{dt'} = X_1' = \frac{1}{l^2\mu}(X_1 + \epsilon\sum X_1\xi) = \frac{l}{l^2\mu}X_1(1 + \epsilon\xi) = \frac{1}{l^2}X_1,$$

$$\varphi(\xi')\frac{d\eta'}{dt'} = Y_1' = \frac{1}{l^2k\mu}Y_1$$

and consequently:

$$\begin{cases} f(\xi)\dfrac{d\xi}{dt} = l^2 f(\xi')\dfrac{d\xi'}{dt'}, \\[2mm] \varphi(\xi)\dfrac{d\eta}{dt} = l^2 k\mu\varphi(\xi')\dfrac{d\eta'}{dt'}. \end{cases} \qquad (11)$$

But we have:

$$\frac{d\xi'}{dt'} = \frac{d\xi}{dt}\frac{1}{k^3\mu^3}, \qquad \frac{d\eta'}{dt'} = \frac{d\eta}{dt}\frac{1}{k^2\mu^2}.$$

hence:

$$f(\xi') = f\left(\frac{\xi + \epsilon}{1 + \xi\epsilon}\right) = f(\xi)\frac{k^3\mu^3}{l^2},$$

$$\varphi(\xi') = \varphi\left(\frac{\xi + \epsilon}{1 + \xi\epsilon}\right) = \varphi(\xi)\frac{k\mu}{l^2},$$

hence, by eliminating the l^2 we find the functional equation:

$$k^2\mu^2\frac{\varphi\left(\frac{\xi+\epsilon}{1+\xi\epsilon}\right)}{\varphi(\xi)} = \frac{f\left(\frac{\xi+\epsilon}{1+\xi\epsilon}\right)}{f(\xi)},$$

or, by posing

$$\frac{\varphi(\xi)}{f(\xi)} = \Omega(\xi) = \frac{D}{V\frac{dD}{dV}},$$

this one

$$\Omega\left(\frac{\xi + \epsilon}{1 + \xi\epsilon}\right) = \Omega(\xi)\frac{1 + \epsilon^2}{(1 + \xi\epsilon)^2},$$

is an equation that must be satisfied for all values in ξ and ϵ. For $\zeta = 0$ we find:

$$\Omega(\epsilon) = \Omega(0)(1 - \epsilon^2),$$

whence:

$$D = A\left(\frac{V}{\sqrt{1 - V^2}}\right)^m,$$

A being a constant and in which I've done $\Omega(0) = \frac{1}{m}$. We then find:

$$\varphi(\xi) = \frac{A}{\xi}\left(\frac{\xi}{\sqrt{1 - \xi^2}}\right)^m, \quad \varphi(\xi') = \frac{A\mu}{\xi + \epsilon}\left(\frac{\xi + \epsilon}{\sqrt{1 - \xi^2}\sqrt{1 - \epsilon^2}}\right)^m.$$

But $\varphi(\xi') = \varphi(\xi)\frac{k\mu}{l^2}$; so we have Since l must depend only on ϵ (since, if there are several electrons, l must be the same for all the electrons whose ξ speeds may be different), this identity can only take place if one has:

$$m = 1, \quad l = 1.$$

Thus the Lorentz hypothesis is the only one that is compatible with the impossibility of demonstrating absolute motion; if we admit this impossibility, we must admit that the electrons in motion contract so as to become ellipsoids of revolution, two of whose axes remaining constant; it is therefore necessary to admit, as we have shown in the previous paragraph, the existence of an additional potential proportional to the volume of the electron.

The analysis of Lorentz is thus fully confirmed, but we can better appreciate the true raison d'être of this analysis; this reason must be sought, among the considerations of paragraph 4. *The transformations that do not alter the equations of motion must form a group, and this is possible only if $l = 1$.* As we must be able to recognize if an electron is at rest or in absolute motion, it is necessary that when it is in motion, it undergoes a deformation which must be precisely that which the corresponding transformation of the group imposes on it.

8. Any given motion

The preceding results apply only to quasi-stationary motion, but it is easy to extend them to the general case; it is sufficient to apply the principles of paragraph 3, i.e. to start from the principle of least action.

To the expression of action

$$J = \int dt d\tau \left(\frac{\sum f^2}{2} - \frac{\sum \alpha^2}{2}\right),$$

a term should be added, representing the additional potential F of paragraph 6; this term will of course take the form:

$$J_1 = \int \sum(F)dt,$$

where $\sum(F)$ represents the sum of the additional potentials due to the different electrons, each of which is proportional to the volume of the corresponding electron.

I write (F) in brackets so that it is not confused with vector F, G, H.

The total action is then $J + J_1$. We have seen in paragraph 3 that J is not altered by the Lorentz transformation; it is necessary to show now that it is the same for J_1.

We have, for one of the electrons,

$$(F) = \omega_0 \tau,$$

ω_0 being a special coefficient of the electron and τ its volume; so I can write:

$$\sum(F) = \int \omega_0 d\tau,$$

the integral must be extended to the whole space, but in such a way that the coefficient ω_0 is null outside the electrons, and inside each electron it is equal to the special coefficient for that electron. We have then:

$$J_1 = \int \omega_0 d\tau dt$$

and after the Lorentz transformation:

$$J_1' = \int \omega_0' d\tau' dt'$$

Now we have $\omega_0 = \omega_0'$; because if a point belongs to an electron, the corresponding point after the Lorentz transformation still belongs to the same electron. On the other hand, we found in paragraph 3;

$$d\tau' dt' = l^4 d\tau dt.$$

and, since we now assume $l = 1$,

$$d\tau' dt' = d\tau dt.$$

So we have

$$J_1 = J_1'.$$

Q.E.D.

The theorem is therefore general, and at the same time it gives us a solution to the question we were asking ourselves at the end of paragraph 1: to find complementary forces not altered by the Lorentz transformation. The additional potential (F) satisfies this condition.

We can therefore generalize the result stated at the end of paragraph 1 and write:

If the inertia of the electrons is exclusively of electromagnetic origin, if they are only subjected to forces of electromagnetic origin, or to the forces that generate the additional potential (F), no experiment will be able to demonstrate absolute motion.

What then are these forces that generate potential (F)? They can obviously be assimilated to a pressure that would prevail inside the electron; everything happens as if each electron were a hollow capacity subjected to a constant internal pressure (independent of volume); the work of such a pressure would obviously be proportional to the variations in volume.

I must observe, however, that this pressure is negative. Let us take again equation (10) of paragraph 6, which in the Lorentz hypothesis is written:

$$F = Ar^3\theta^2;$$

equations (11) of paragraph 6 will give us:

$$A = \frac{a}{3b^4}.$$

Our pressure is equal to A, except for a constant coefficient, which is negative.

Let's now evaluate the mass of the electron, I mean the "experimental mass," i.e. mass for low velocities; we have (cf. paragraph 6):

$$H = \frac{\varphi\left(\frac{\theta}{k}\right)}{k^2 r}, \quad \theta = k, \quad \varphi = a, \quad \theta r = b;$$

whence

$$H = \frac{a}{bk} = \frac{a}{b}\sqrt{1 - V^2},$$

For V very small, I can write:

$$H = \frac{a}{b}\left(1 - \frac{V^2}{2}\right),$$

so that the mass, both longitudinal and transverse, will be

Now a is a numerical constant, which shows that: *the pressure that generates our additional potential is proportional to the fourth power of the experimental mass of the electron.*

Since the Newtonian attraction is proportional to this experimental mass, one is tempted to conclude that there is some relationship between the cause that generates gravitation and the cause that generates this extra potential.

9. Hypotheses about Gravitation

Thus the Lorentz theory would completely explain the impossibility to demonstrate absolute motion, if all forces were of electromagnetic origin.

But there are forces that cannot be attributed to an electromagnetic origin, such as gravitation, for example. It can happen, in fact, that two systems of bodies produce equivalent electromagnetic fields, i.e. exerting the same action on electrified bodies and on currents, and yet that these two systems do not exert the same gravitational action on Newtonian masses. The gravitational field is therefore distinct from the electromagnetic field. Lorentz was therefore obliged to complete his hypothesis by assuming *that forces of any origin, and in particular gravitation, are affected by a translation (or, if one prefers, by the Lorentz transformation) in the same way as electromagnetic forces.*

It is now appropriate to go into detail and examine this hypothesis in more detail. If we want the Newtonian force to be affected in this way by the Lorentz transformation, we can no longer allow that this force depends solely on the relative position of the attracting body and the attracted body at the moment considered. It should also depend on the velocities of the two bodies. And that's not all: it will be natural to assume that the force that acts at instant t on the attracted body, depends on the position and velocity of this body at this same instant t; but it will depend, moreover, on the position and velocity of the *attracting* body, not at instant t, but at an instant earlier, as if gravity had taken some time to propagate.

Let us thus consider the position of the attracted body at instant t_0 and let, at this instant, x_0, y_0, z_0 be its coordinates, ξ, η, ζ, the components of its velocity; let us consider on the other hand the attracting body at the corresponding instant $t_0 + t$ and let, at this instant, $x_0 + x, y_0 + y, z_0 + z$ be its coordinates, $\xi_1, \eta_1, zeta_1$ the components of its velocity. We must first have a relation

$$\varphi(t, x, y, z, \xi, \eta, \zeta, \xi_1, \eta_1, \zeta_1) = 0 \tag{1}$$

to define time t. This relation will define the law of propagation of the gravitational action (I do not impose on myself the condition that the propagation is done with the same velocity in all directions).

Let now X_1, Y_1, Z_1 be the 3 components of the action exerted at the instant t_0 on the attracted body; this consists in expressing X_1, Y_1, Z_1 as functions of

$$t, x, y, z, \xi, \eta, \zeta, \xi_1, \eta_1, \zeta_1. \tag{2}$$

What are the conditions to be filled?

1^0 Condition (1) must not be altered by the transformations of the Lorentz group.

2^0 The components X1, Y1, Z1 shall be affected by the Lorentz transformations in the same way as the electromagnetic forces identified by the same letters, i.e. in accordance with equations (11 bis) of paragraph 1.

3^0 When the two bodies are at rest, we must return to the ordinary law of attraction.

It is important to note that in the latter case, relation (1) disappears, because time t no longer plays any role if both bodies are at rest.

The problem thus posed is obviously undetermined. We will therefore seek to satisfy as much as possible other complementary conditions:

4^0 Since astronomical observations apparently do not show any significant deviation from Newton's law, we will choose the solution that least deviates from this law for low velocities of both bodies.

5^0 We will try to manage so that t is always negative; indeed, if we conceive that the effect of gravitation requires a certain time to propagate, it would be more difficult to understand how this effect could depend on the position *not yet reached* by the attracting body.

There is a case where the indeterminacy of the problem disappears; it is the case in which both bodies are at rest *relative* to each other, that is, where:

$$\xi = \xi_1, \quad \eta = \eta_1, \quad \zeta = \zeta_1;$$

So this is the case that we will examine first, assuming that these velocities are constant, so that both bodies are driven in a common, rectilinear and uniform translational motion.

We can assume that the x-axis has been taken parallel to this translation, so that $\eta = \zeta = 0$, and we will take $\epsilon = -\xi$.

If under these conditions we apply the Lorentz transformation, after the transformation both bodies will be at rest and we will have:

$$\xi' = \eta' = \zeta' = 0.$$

Then the components X_0', Y_0', Z_0' will have to conform to Newton's law and we will have, to within a constant factor:

$$\begin{cases} X_1' = -\dfrac{x'}{r'^3}, \quad Y_1' = -\dfrac{y'}{r'^3}, \quad Z_1' = -\dfrac{z'}{r'^3}, \\ r'^2 = x'^2 + y'^2 + z'^2. \end{cases} \tag{3}$$

But we have, according to paragraph 1:

$$\begin{cases} x' = k(x + \epsilon t), \quad y' = y, \quad z' = z, \quad t' = k(t + \epsilon x), \\ \dfrac{\rho'}{\rho} = h(1 - \xi\epsilon) = k(1 - \epsilon^2) = \dfrac{1}{k}, \quad \sum X_1 \xi = -X_1 \epsilon, \\ X_1' = k\dfrac{\rho}{rho'}\left(X_1 + \epsilon \sum X_1 \xi\right) = k^2 X_1(1 - \epsilon^2) = X_1, \\ Y_1' = \dfrac{\rho}{\rho'} Y_1 = kY_1, \\ Z_1' = kZ_1. \end{cases}$$

We have besides:

$$x + \epsilon t = x - \xi t, \quad r'^2 = k^2(x - \xi t)^2 + y^2 + z^2$$

and

$$X_1 = -\frac{k(x - \xi t)}{r'^3}, \quad Y_1 = -\frac{y}{kr'^3}, \quad Z_1 = -\frac{z}{kr'^3}; \tag{4}$$

which can be written:

$$X_1 = \frac{dV}{dx}, \quad Y_1 = \frac{dV}{dy}, \quad Z_1 = \frac{dV}{dz}; \quad V = \frac{1}{kr'}. \tag{4 bis}$$

It seems first that the indeterminacy remains, since we have made no hypothesis on the value of t, i.e. on the velocity of the transmission; and that moreover x is a function of t; but it is easy to see that $x - \epsilon t, y, z$, that appear by themselves in our formulas, do not depend on t.

We see that if the two bodies are simply animated by a common translation, the force that acts on the attracted body is normal to an ellipsoid with the attractive body as its centre.

To go further it is necessary to look for the *invariants of the Lorentz group*.

We know that the substitutions in this group (assuming $l = 1$) are the linear substitutions that do not alter the quadratic form

$$x^2 + y^2 + z^2 - t^2.$$

Let's pose on the other hand:

$$\xi = \frac{\delta x}{\delta t}, \quad \eta = \frac{\delta y}{\delta t}, \quad \zeta = \frac{\delta z}{\delta t};$$

$$\xi_1 = \frac{\delta_1 x}{\delta_1 t}, \quad \eta_1 = \frac{\delta_1 y}{\delta_1 t}, \quad \zeta_1 = \frac{\delta_1 z}{\delta_1 t};$$

we see that the Lorentz transformation will have the effect of causing δx, δy, δz, δt and $\delta_1 x$, $\delta_1 y$, $\delta_1 z$, $\delta_1 t$ to undergo the same linear substitutions as x, y, z, t.

Let us consider

$$
\begin{array}{cccc}
x, & y, & z, & t\sqrt{-1}, \\
\delta x, & \delta y, & \delta z, & \delta t\sqrt{-1}, \\
\delta_1 x, & \delta_1 y, & \delta_1 z, & \delta_1 t\sqrt{-1},
\end{array}
$$

as the coordinates of three points P, P', P'' in four-dimensional space. We see that the Lorentz transformation is only a rotation of this space around the origin, regarded as fixed. We will therefore have no distinct invariants other than the 6 distances of the 3 points P, P', P'' from each other and from the origin, or, if one prefers, only the two expressions:

$$x^2 + y^2 + z^2 - t^2, \quad x\delta x + y\delta y + z\delta z - t\delta t,$$

or the four expressions of the same form that can be deduced by swapping in any way the three points P, P', P''.

But what we are looking for are the functions of the ten variables (2) which are invariants; we must therefore, among the combinations of our six invariants, look for those that depend only on these ten variables, i.e. those that are homogeneous of degree 0 both with respect to $\delta x, \delta y, \delta z, \delta t$ and with respect to $\delta_1 x, \delta_1 y, \delta_1 z, \delta_1 t$. We will thus be left with four distinct invariants, which are:

$$\sum x^2 - t^2, \quad \frac{t - \sum x\xi}{\sqrt{1 - \sum \xi^2}}, \quad \frac{t - \sum x\xi_1}{\sqrt{1 - \sum \xi_1^2}}, \quad \frac{1 - \sum \xi\xi_1}{\sqrt{(1 - \sum \xi^2)(1 - \sum \xi_1^2)}}. \quad (5)$$

Let us now deal with the transformations undergone by the components of force; let us take up again equations (11) of paragraph 1, that relate not to force $X_1, Y_1, Z_1,$

that we consider here, but to force X, Y, Z related to the unit of volume. So let us posit:

$$T = \sum X\xi;$$

we will see that these equations (11) can be written $(l = 1)$

$$\begin{cases} X' = k(X = \epsilon T), & T' = k(T + \epsilon X), \\ Y' = Y, & Z' = Z \end{cases} \tag{6}$$

so that X, Y, Z, T undergo the same transformation as x, y, z, t. The invariants of the group will therefore be

$$\sum X^2 - T^2, \quad \sum Xx - Tt, \quad \sum X\delta x - T\delta t, \quad \sum X\delta_1 x - T\delta_1 t.$$

But it's not X, Y, Z that we need, but X_1, Y_1, Z_1 with

$$T_1 = \sum X_1 \xi.$$

We see that

$$\frac{X_1}{X} = \frac{Y_1}{Y} = \frac{Z_1}{Z} = \frac{T_1}{T} = \frac{1}{\rho}.$$

So the Lorentz transformation will act on X_1, Y_1, Z_1, T_1 in the same way as on X, Y, Z, T, with the difference that these expressions will be moreover multiplied by

$$\frac{\rho}{\rho'} = \frac{1}{k(1 - \xi\epsilon)} = \frac{\delta t}{\delta t'}.$$

Likewise it will act on $\xi, \eta, \zeta, 1$, in the same way as on $\delta x, \delta y, \delta z, \delta t$, with the difference that these expressions will also be multiplied by the *same* factor:

$$\frac{\delta t}{\delta t'} = \frac{1}{k(1 + \xi\epsilon)}.$$

Let us consider $X, Y, Z, T\sqrt{-1}$, as the coordinates of a fourth point Q; then the invariants will be of the mutial distances of the five points

$$0, \quad P, \quad P', \quad P'', \quad Q$$

and among these functions we must keep only those that are homogeneous of degree 0, on the one hand in relation to

$$X, \quad Y, \quad Z, \quad T, \quad \delta x, \quad \delta y, \quad \delta z, \quad \delta t$$

(variables that can be replaced by $X_1, Y_1, Z_1, T_1, \xi, \eta, \zeta, 1$), on the other hand with respect to

$$\delta_1 x, \quad \delta_1 y, \quad \delta_1 z, \quad 1$$

(variables that can then be replaced by $\xi_1, \eta_1, \zeta_1, 1$). We thus find, in addition to the four invariants (5), four distinct new invariants, which are:

$$\frac{\sum X_1^2 - T_1^2}{1 - \sum \xi^2}, \quad \frac{\sum X_1 x - T_1 t}{\sqrt{1 - \sum \xi^2}}, \quad \frac{\sum X_1 \xi_1 - T_1}{\sqrt{1 - \sum \xi^2}\sqrt{1 - \sum \xi_1^2}}, \quad \frac{\sum X_1 \xi - T_1}{1 - \sum \xi^2}. \quad (7)$$

The last invariant is always zero, according to the definition of T_1.

This having been said, what are the conditions to be met?

1^0 The first member of the relation (1), that defines the propagation rate, must be a function of the four invariants (5).

Numerous hypotheses can of course be made; we will examine only two of them:

a. We can have

$$\sum x^2 - t^2 = r^2 - t^2 = 0, \quad \text{whence} \quad t = \pm r,$$

and, since t must be negative, $t = -r$. This means that the propagation velocity is equal to that of light. It seems first of all that this hypothesis must be rejected without examination. Laplace has indeed shown that this propagation is either instantaneous or much faster than that of light. But Laplace had examined the hypothesis of the finite speed of propagation, *ceteris non mutatis;* here, contrariwise, this hypothesis is complicated by many others, and it may be that there is a more or less perfect compensation between them, such as those of which the applications of the Lorentz transformation have already given us so many examples.

b. We can have

$$\frac{1 - \sum x\xi_1}{\sqrt{1 - \sum \xi_1^2}} = 0, \quad t = \sum x\xi_1.$$

The propagation velocity is then much faster than that of light; but in some cases t could be negative, which, as we have said, does not seem to be acceptable. *We will therefore stick to hypothesis (a).*

2^0 The four invariants (7) must be functions of the invariants (5).

3^0 When both bodies are at absolute rest, X_1, Y_1, Z_1 must have the value deduced from Newton's law, and when they are at relative rest, the value deduced from equations (4).

Under the hypothesis of absolute rest, the first two invariants (7) must be reduced to

$$\sum X_1^2, \quad \sum X_1 x,$$

or, by Newton's law, to

$$\frac{1}{r^4}, \quad -\frac{1}{r};$$

on the other hand, in hypothesis (a), the second and third of the invariants (5) become:

$$\frac{-r - \sum x\xi}{\sqrt{1 - \sum \xi^2}}, \quad \frac{-r - \sum x\xi_1}{\sqrt{1 - \sum \xi_1^2}},$$

that is, for absolute rest, at

$$- r, \quad -r.$$

We can therefore assume, *for example*, that the first two invariants (4) are reduced to

$$\frac{(1 - \sum \xi_1^2)^2}{(r + \sum x\xi_1)^4}, \quad -\frac{\sqrt{1 - \sum \xi_1^2}}{r + \sum x\xi_1},$$

But other combinations are possible.

We have to make a choice between these combinations, and, on the other hand, to define X_1, Y_1, Z_1, we need a third equation. For such a choice, we must strive to come as close as possible to Newton's law. So let's see what happens when (always using $t = -r$) we neglect the squares of velocities ξ, η and so on. The four invariants (5) become then:

$$0, \quad -r - \sum x\xi, \quad -r - \sum x\xi_1, \quad 1$$

and the four invariants (7):

$$\sum X_1^2, \quad \sum X_1(x + \xi r), \quad \sum X_1(\xi_1 - \xi), \quad 0.$$

But to be able to compare with Newton's law, another transformation is necessary; here $x_0 + x, y_0 + y, z_0 + z$ represents the coordinates of the attracting body at instant $t_0 + t$ and $r = \sqrt{\sum x^2}$; in Newton's law it is necessary to consider the coordinates $x_0 + x_1, y_0 + y_1, z_0 + z_1$ of the attracting body at instant t_0 and distance $r_1 = \sqrt{\sum x_1^2}$.

We can neglect the square of time t required for propagation and therefore assume that the motion is uniform; we have then:

$$x = x_1 + \xi_1 t, \quad y = y_1 + \eta_1 t, \quad z = z_1 + \zeta_1 t, \quad r(r - r_1) = \sum x\xi_1 t,$$

or, since $t = -r$,

$$x = x_1 - \xi_1 r, \quad y = y_1 - \eta_1 r, \quad z = z_1 - \zeta_1 r, \quad r = r_1 - \sum x\xi_1;$$

so that our four invariants (5) become:

$$0, \quad -r_1 + \sum x(\xi_1 - \xi), \quad -r_1, \quad 1$$

and our four invariants (7):

$$\sum X_1^2, \quad \sum X_1[x_1 + (\xi - \xi_1)r_1], \quad \sum X_1(\xi_1 - \xi), \quad 0.$$

In the second of these expressions I wrote r_1 instead of r, because r is multiplied by $\xi - \xi_1$ and I neglect the square of ξ.

On the other hand, Newton's law would give us, for these four invariants (7),

$$\frac{1}{r_1^4}, \quad -\frac{1}{r_1} - \frac{\sum x_1(\xi - \xi_1)}{r_1^2}, \quad \frac{\sum x_1(\xi - \xi_1)}{r_1^3}, \quad 0.$$

If then we call A and B, the second and third of the invariants (5), and M, N, P the first three invariants (7), we will satisfy Newton's law, near the terms of the square of the velocities, by posing:

$$M = \frac{1}{B^4}, \quad N = +\frac{A}{B^2}, \quad P = \frac{A - B}{B^3}. \tag{8}$$

This solution is not unique. Let C be the fourth invariant (5), $C - 1$ is of the order of the square of ξ, and so is $(A - B)^2$.

We could therefore add to the second members of each of equations (8) a term consisting of $C-1$ multiplied by an arbitrary function of A, B, C and a term consisting of $(A - B)^2$ also multiplied by a function of A, B, C.

At first glance, the solution (8) seems the simplest, but it cannot be adopted; in fact, since M, N, P are functions of X_1, Y_1, Z_1 and $T_1 = \sum X_1\xi$, we can derive from these three equations (8) the values of X_1, Y_1, Z_1; but in some cases these values would become imaginary.

To avoid this inconvenience, we will operate differently. Let us pose:

$$k_0 = \frac{1}{\sqrt{1 - \sum \xi^2}}, \quad k_1 = \frac{1}{\sqrt{1 - \sum \xi_1^2}},$$

which is justified by analogy with the notation

$$k = \frac{1}{\sqrt{1 - \epsilon^2}}$$

that appears in the Lorentz transformation.

In this case, and due to condition $t = -r$, the invariants (5) become:

$$0, \quad A = -k_0\left(r + \sum x\xi\right), \quad B = -k_1\left(r - \sum x\xi_1\right), \quad C = k_0 k_1\left(1 - \sum \xi\xi_1\right).$$

On the other hand, we see that the following systems of quantities:

x	y	z	$-r = t$
$k_0 X_1$	$k_0 Y_1$	$k_0 Z_1$	$k_0 T_1$
$k_0 \xi$	$k_0 \eta$	$k_0 \zeta$	k_0
$k_1 \xi_1$	$k_1 \eta_1$	$k_1 \zeta_1$	k_1

undergo the *same* linear substitutions when the transformations of the Lorentz group are applied to them. We are therefore led to pose:

$$\begin{cases} X_1 = x\dfrac{\alpha}{k_0} + \xi\beta + \xi_1\dfrac{k_1}{k_0}\gamma, \\[2mm] Y_1 = y\dfrac{\alpha}{k_0} + \eta\beta + \eta_1\dfrac{k_1}{k_0}\gamma, \\[2mm] Z_1 = z\dfrac{\alpha}{k_0} + \zeta\beta + \zeta_1\dfrac{k_1}{k_0}\gamma, \\[2mm] T_1 = -r\dfrac{\alpha}{k_0} + \beta + \dfrac{k_1}{k_0}\gamma. \end{cases} \tag{9}$$

It is clear that if α, β, γ are invariants, X_1, Y_1, Z_1, T_1 will satisfy the fundamental condition, i.e. they will undergo, through the effect of Lorentz transformations, a suitable linear substitution. But for equations (9) to be compatible, we must have:

$$\sum X_1\xi - T_1 = 0,$$

which, by replacing X_1, Y_1, Z_1, T_1 with their values (9) and multiplying by k_0^2, becomes:

$$- A\alpha - \beta - C\gamma = 0. \tag{10}$$

What we want is that if we neglect, in front of the square of the speed of light, the squares of speeds ξ etc., as well as the product of accelerations by distances, as we have done above, the values of X_1, Y_1, Z_1 remain in accordance with Newton's law.

We can take:

$$\beta = 0, \quad \gamma = -\frac{A\alpha}{C}.$$

With the order of approximation adopted, we have:

$$k_0 = k_1 = 1, \quad C = 1, \quad A = -r_1 + \sum x(\xi_1 - \xi), \quad B = -r_1,$$

$$x = x_1 + \xi_1 t = x_1 - \xi_1 r.$$

The first equation (9) then becomes:

$$X_1 = \alpha(x - A\xi_1).$$

But if we neglect the square of ξ, we can replace $A\xi_1$ by $-r_1\xi_1$ or by $r\xi_1$ which gives:

$$X_1 = \alpha(x + \xi_1 r) = \alpha x_1.$$

Newton's law would give

$$X_1 = -\frac{x_1}{r_1^3}.$$

We must therefore choose, for the invariant α, the one that is reduced to $-\frac{1}{r_1^3}$ to the adopted order of approximations, i. e., $\frac{1}{B^3}$. The equations (9) become:

$$
\begin{cases}
X1 = \dfrac{x}{k_0 B^3} - \xi_1 \dfrac{k_1}{k_0} \dfrac{A}{B^3 C}, \\[2mm]
Y1 = \dfrac{y}{k_0 B^3} - \eta_1 \dfrac{k_1}{k_0} \dfrac{A}{B^3 C}, \\[2mm]
Z1 = \dfrac{z}{k_0 B^3} - \zeta_1 \dfrac{k_1}{k_0} \dfrac{A}{B^3 C}, \\[2mm]
T1 = -\dfrac{r}{k_0 B^3} - \dfrac{k_1}{k_0} \dfrac{A}{B^3 C},
\end{cases}
\tag{11}
$$

We first see that the corrected attraction is composed of two components; one parallel to the vector that joins the positions of the two bodies, the other parallel to the velocity of the attracting body.

Let's remember that when we talk about the position or velocity of the attracting body, we are talking about its position or velocity at the moment when the gravitational wave leaves it; for the attracted body, on the contrary, it is its position or velocity at the moment when the gravitational wave reaches it, this wave being assumed to propagate with the speed of light.

I think it would be premature to discuss these formulas any further, so I will restrict myself to a few remarks.

1^0 Solutions (11) are not unique; in fact, one can replace $\frac{1}{B^3}$, which is a factor everywhere, by

$$
\frac{1}{B^3} + (C - 1)f_1(A, B, C) + (A - B)^2 f_2(A, B, C),
$$

f_1 and f_2 being arbitrary functions of A, B, C, or no longer taking β as null and void, but adding to α, β, γ any additional terms, provided that they satisfy condition (10) and that they are of the second order with respect to ξ, as far as α is concerned, and of the first order with respect to β and γ.

2^0 The first equation (11) can be written:

$$
X_1 = \frac{k_1}{B^3 C}\left[x(1 - \sum \xi\xi_1) + \xi_1(r + \sum x\xi)\right]
\tag{11 bis}
$$

and the amount in square brackets can, itself, be written:

$$
(x + r\xi_1) + \eta(\xi_1 y - x\eta_1) + \zeta(\xi_1 z - x\zeta_1),
\tag{12}
$$

so that the total force can be divided into three components corresponding to the three parentheses of expression (12); the first component has a vague analogy with the mechanical force due to the electric field, the other two with the mechanical force due to the magnetic field; to complete the analogy I can, by virtue of the first remark, replace in equations (11) $\frac{1}{B^3}$ by $\frac{C}{B^3}$, so that X_1, Y_1, Z_1 depend only linearly on the velocity ξ, η, ζ of the attracted body, since C has disappeared from the denominator of (11 bis).

Let us pose then:

$$\begin{cases} k_1(x + r\xi_1) = \lambda, \quad k_1(y + r\eta_1) = \mu, \quad k_1(z + r\zeta_1) = \nu, \\ k_1(\eta_1 z - \zeta_1 y) = \lambda', \quad k_1(\zeta_1 x - \xi_1 z) = \mu', \quad k_1(\xi_1 y - x\eta_1) = \nu'; \end{cases} \tag{13}$$

we will have, C having disappeared from the denominator of (11 bis):

$$\begin{cases} X_1 = \dfrac{\lambda}{B^3} + \dfrac{\eta\nu' - \zeta\mu'}{B^3} \\ Y_1 = \dfrac{\mu}{B^3} + \dfrac{\zeta\lambda' - \xi\nu'}{B^3} \\ Z_1 = \dfrac{\nu}{B^3} + \dfrac{\xi\mu' - \eta\lambda'}{B^3}, \end{cases} \tag{14}$$

and we'll have, besides:

$$B^2 = \sum \lambda^2 - \sum \lambda'^2. \tag{15}$$

So λ, μ, ν, where $\frac{\lambda}{B^3}, \frac{\mu}{B^3}, \frac{\nu}{B^3}$ is a kind of electric field, while λ', μ', ν' or rather $\frac{\lambda'}{B^3}, \frac{\mu'}{B^3}, \frac{\nu'}{B^3}$ is a kind of magnetic field.

3^0 The postulate of relativity would compel us to adopt solution (11) or solution (14) or any of the solutions that would be deduced from the first remark; but the first question that arises is whether they are compatible with astronomical observations; the divergence with Newton's law is of the order of ξ^2, i.e. $10,000$ times smaller than if it were of the order of ξ, i.e. if the propagation were made with the speed of light, *ceteris non mutatis*; it is therefore to be hoped that it will not be too great. But only a thorough discussion will tell us.

Paris, July 1905.   H. POINCARÉ

ON THE ELECTRODYNAMICS OF MOVING BODIES

A. EINSTEIN

Translation of Zur Elektrodynamik bewegter Körper, *Annalen der Physik*, 17, 1905. Original English publication in: H. A. Lorentz, A. Einstein, H. Minkowski and H. Weyl, *The Principle of Relativity: A Collection of Original Memoirs on the Special and General Theory of Relativity*. With Notes by A. Sommerfeld. Translated by W. Perrett and G. B. Jeffery (Methuen and Company, Ltd., 1923; reprinted by Dover Publications Inc., 1952).

It is known that Maxwell's electrodynamics – as usually understood at the present time – when applied to moving bodies, leads to asymmetries which do not appear to be inherent in the phenomena. Take, for example, the reciprocal electrodynamic action of a magnet and a conductor. The observable phenomenon here depends only on the relative motion of the conductor and the magnet, whereas the customary view draws a sharp distinction between the two cases in which either the one or the other of these bodies is in motion. For if the magnet is in motion and the conductor at rest, there arises in the neighbourhood of the magnet an electric field with a certain definite energy, producing a current at the places where, parts of the conductor are situated. But if the magnet is stationary and the conductor in motion, no electric field arises in the neighbourhood of the magnet. In the conductor, however, we find an electromotive force, to which in itself there is no corresponding energy, but which gives rise – assuming equality of relative motion in the two cases discussed – to electric currents of the same path and intensity as those produced by the electric forces in the former case.

Examples of this sort, together with the unsuccessful attempts to discover any motion of the earth relatively to the "light medium," suggest that the phenomena of electrodynamics as well as of mechanics possess no properties corresponding to the idea of absolute rest.[1] They suggest rather that, as has already been shown to the

---

[1] EDITOR'S NOTE: There has been some confusion on whether Einstein was aware of the Michelson-Morley experiment when he wrote this (1905) paper. It seems what what might have contributed to the confusion was an inappropriate and unfortunate change of a phrase in Minkowski's lecture "The Relativity Principle" (given at the meeting of the Göttingen Mathematical Society on November 5, 1907) made by Sommerfeld when, after Minkowski's death in January 1909, he prepared Minkowski's lecture for publication in 1915 – as Pyenson (L. Pyenson, Hermann Minkowski and Einstein's Special Theory of Relativity, *Archive for History of Exact Sciences* 17 (1977) pp. 71-95, p. 82) explained: "Sommerfeld was unable to resist rewriting Minkowski's judgment of Einstein's

first order of small quantities, the same laws of electrodynamics and optics will be valid for all frames of reference for which the equations of mechanics hold good.[2] We will raise this conjecture (the purport of which will hereafter be called the "Principle of Relativity") to the status of a postulate, and also introduce another postulate, which is only apparently irreconcilable with the former, namely, that light is always propagated in empty space with a definite velocity c which is independent of the state of motion of the emitting body. These two postulates suffice for the attainment of a simple and consistent theory of the electrodynamics of moving bodies based on Maxwell's theory for stationary bodies. The introduction of a "luminiferous ether" will prove to be superfluous inasmuch as the view here to be developed will not require an "absolutely stationary space" provided with special properties, nor assign a velocity-vector to a point of the empty space in which electromagnetic processes take place.

The theory to be developed is based –like all electrodynamics – on the kinematics of the rigid body, since the assertions of any such theory have to do with the relationships between rigid bodies (systems of coordinates), clocks, and electromagnetic processes. Insufficient consideration of this circumstance lies at the root of the difficulties which the electrodynamics of moving bodies at present encounters.

I. KINEMATICAL PART

§ 1. Definition of Simultaneity

Let us take a system of coordinates in which the equations of Newtonian mechanics hold good.[3] In order to render our presentation more precise and to distinguish this system of coordinates verbally from others which will be introduced hereafter, we call it the "stationary system."

---

formulation of the principle of relativity. He introduced a clause inappropriately praising Einstein for having used the Michelson experiment to demonstrate that the concept of absolute space did not express a property of phenomena. Sommerfeld also suppressed Minkowski's conclusion,where Einstein was portrayed as the clarifier, but by no means as the principal expositor, of the principle of relativity." Here is the altered text: "particularly in light of *Michelson*'s experiment, it has been shown that, as Einstein so succinctly expresses this, the concept of an absolute state of rest entails no properties that correspond to phenomena" ("The Relativity Principle" in: Hermann Minkowski, *Spacetime: Minkowski's Papers on Spacetime Physics* (Minkowski Institute Press, Montreal 2020), p. 82). Giving credit to Einstein for realizing the crucial role of the Michelson experiment is especially unfortunate since Einstein himself stated the opposite: "In my own development, Michelson's result has not had a considerable influence. I even do not remember if I knew of it at all when I wrote my first paper on the subject (1905). The explanation is that I was, for general reasons, firmly convinced that there does not exist absolute motion and my problem was only how this could be reconciled with our knowledge of electrodynamics. One can therefore understand why in my personal struggle Michelson's experiment played no role, or at least no decisive role." (A. Pais, *Subtle Is the Lord: The Science and the Life of Albert Einstein* (Oxford University Press, Oxford 2005) p. 172). However, the clearest indication that Einstein had been unaware of the Michelson-Morley experiment is the phrase (in the next sentence above) "as has already been shown for quantities of the first order." The Michelson-Morley experiment deals with quantities of the second order.

[2]The preceding memoir by Lorentz (the second paper by Lorentz in this volume) was not at this time known to the author.

[3]I.e. to the first approximation.

If a material point is at rest relatively to this system of coordinates, its position can be defined relatively thereto by the employment of rigid standards of measurement and the methods of Euclidean geometry, and can be expressed in Cartesian coordinates.

If we wish to describe the *motion* of a material point, we give the values of its coordinates as functions of the time. Now we must bear carefully in mind that a mathematical description of this kind has no physical meaning unless we are quite clear as to what we understand by "time." We have to take into account that all our judgments in which time plays a part are always judgments of *simultaneous events*. If, for instance, I say, "That train arrives here at 7 o'clock," I mean something like this: "The pointing of the small hand of my watch to 7 and the arrival of the train are simultaneous events."[4]

It might appear possible to overcome all the difficulties attending the definition of "time" by substituting "the position of the small hand of my watch" for "time." And in fact such a definition is satisfactory when we are concerned with defining a time exclusively for the place where the watch is located; but it is no longer satisfactory when we have to connect in time series of events occurring at different places, or − what comes to the same thing − to evaluate the times of events occurring at places remote from the watch.

We might, of course, content ourselves with time values determined by an observer stationed together with the watch at the origin of the coordinates, and coordinating the corresponding positions of the hands with light signals, given out by every event to be timed, and reaching him through empty space. But this coordination has the disadvantage that it is not independent of the standpoint of the observer with the watch or clock, as we know from experience. We arrive at a much more practical determination along the following line of thought.

If at the point A of space there is a clock, an observer at A can determine the time values of events in the immediate proximity of A by finding the positions of the hands which are simultaneous with these events. If there is at the point B of space another clock in all respects resembling the one at A, it is possible for an observer at B to determine the time values of events in the immediate neighbourhood of B. But it is not possible without further assumption to compare, in respect of time, an event at A with an event at B. We have so far defined only an "A time" and a "B time." We have not defined a common "time" for A and B, for the latter cannot be defined at all unless we establish *by definition* that the "time" required by light to travel from A to B equals the "time" it requires to travel from B to A.[5] Let a ray of light start

---

[4]We shall not here discuss the inexactitude which lurks in the concept of simultaneity of two events at approximately the same place, which can only be removed by an abstraction.

[5]EDITOR'S NOTE: This sentence, and especially the phrase "*by definition*" gave rise to a long-running debate on whether the one-way velocity of light and simultaneity are a matter of *convention*. Indeed, Einstein's assertion "we establish *by definition* that the "time" required by light to travel from A to B equals the "time" it requires to travel from B to A" means that "we establish *by definition*" that the velocity of light from A to B and from B to A is the *same* and therefore the synchronization of the clocks at A and B (i.e., the *simultaneity* of the readings of the clocks at A and B) is also established *by definition*.

Often the instant temptation, after reading Einstein's assertion, is to ask "Why by definition? One can simply measure the one-way velocities of light from A to B and from B to A and determine

at the "*A* time" t_A from *A* towards *B*, let it at the "*B* time" t_B be reflected at *B* in the direction of *A*, and arrive again at *A* at the "*A* time" t'_A. By definition, the two clocks run synchronously if

$$t_B - t_A = t'_A - t_B$$

We assume that this definition of synchronism is free from contradictions, and possible for any number of points; and that the following relations are universally valid:

1. If the clock at *B* synchronizes with the clock at *A*, the clock at *A* synchronizes with the clock at *B*.

2. If the clock at *A* synchronizes with the clock at *B* and also with the clock at *C*, the clocks at *B* and *C* also synchronize with each other.

Thus with the help of certain imaginary physical experiments we have settled what is to be understood by synchronous stationary clocks located at different places, and have evidently obtained a definition of "simultaneous," or "synchronous," and of "time." The "time" of an event is that which is given simultaneously with the event

---

experimentally whether they are the same." But that instant temptation immediately evaporates when the next self-evident question is asked: "Was Einstein that dumb to fail to ask himself the above exceedingly obvious question?" A single fact makes it instantly clear that not only did Einstein certainly ask that question but he somehow realized that the one-way velocity of light could not be determined by experiment – at the time of the publication of this paper in 1905 he had been still working in the patent office in Bern, where he had examined many patent applications with proposals on how to synchronize the clocks of train stations.

It is unclear how, by analyzing the procedure of synchronizing distant clocks, Einstein arrived at the conclusion that both the one-way velocity of light and (therefore) the simultaneity of distant events could not be determined by experiment and are therefore a matter of convention. Apparently Einstein merely *postulated*, as he did with the principle of relativity and the constancy of the velocity of light (which Minkowski *explained*), that both the one-way velocity of light and (therefore) the simultaneity of distant events could be determined only *by definition*.

However, to try to reconstruct Einstein's analysis will be a rewarding effort – such an analysis demonstrates that any attempt to synchronize distant clocks (by sending light signals or by slow transport of a third clock) inescapably leads to a *vicious circle*. It seems certain that Einstein had realized this problem. Had he abandoned his approach to postulate things and had asked what the *physical meaning* of that vicious circle was, he would have found a different path to the discovery of the spacetime structure of the world (and to the concept of spacetime). For details, see V. Petkov, "Conventionality of Simultaneity and Reality." In: D. Dieks (ed.), *The Ontology of Spacetime II* (Elsevier, Amsterdam 2008); "Philosophy and Foundations of Physics" Series, Volume 4, pp. 175-185 and V. Petkov, *Relativity and the Nature of Spacetime*, 2nd ed. (Springer, Heidelberg 2009) Sec. 6.1.

It should be noted that the debate over the conventionality issue is a result of not taking the spacetime structure of the world, introduced by Minkowski, seriously. There is no absolute (not conventional) simultaneity, because due to the equal existential status of the events of spacetime there does not exist a class of events that are objectively simultaneous (rigorously speaking, absolute simultaneity can exist only in a three-dimensional world, because such a world and a three-dimensional space are defined in terms of *simultaneity* – as the class of simultaneous events at a given moment of time). The assignment of a class of simultaneous events to a reference frame is indeed "*by definition*" because it is merely a *description* in terms of our everyday three-dimensional language of the monolithic spacetime, in which (i.e., in the physical world) there is no such thing as a class of simultaneous events. That the one-way velocity of light is also a matter of convention can be understood even better if it is taken into account that there is no such thing as velocity of light (and velocity of anything) in spacetime; there are only worldlines in spacetime – light-like and time-like – representing what we perceive as *motion* of light and particles, respectively.

by a stationary clock located at the place of the event, this clock being synchronous, and indeed synchronous for all time determinations, with a specified stationary clock.

In agreement with experience we further assume the quantity

$$\frac{2AB}{t'_A - t_A} = c,$$

to be a universal constant – the velocity of light in empty space.

It is essential to have time defined by means of stationary clocks in the stationary system, and the time now defined being appropriate to the stationary system we call it "the time of the stationary system."

§ 2. On the Relativity of Lengths and Times

The following reflections are based on the principle of relativity and on the principle of the constancy of the velocity of light. These two principles we define as follows:

1. The laws by which the states of physical systems undergo change are not affected, whether these changes of state be referred to the one or the other of two systems of coordinates in uniform translatory motion.

2. Any ray of light moves in the "stationary" system of coordinates with the determined velocity c, whether the ray be emitted by a stationary or by a moving body. Hence

$$\text{velocity} = \frac{\text{light path}}{\text{time interval}}$$

where time interval is to be taken in the sense of the definition in § 1.

Let there be given a stationary rigid rod; and let its length be l as measured by a measuring-rod which is also stationary. We now imagine the axis of the rod lying along the axis of x of the stationary system of coordinates, and that a uniform motion of parallel translation with velocity v along the axis of x in the direction of increasing x is then imparted to the rod. We now inquire as to the length of the moving rod, and imagine its length to be ascertained by the following two operations:

a) The observer moves together with the given measuring-rod and the rod to be measured, and measures the length of the rod directly by superposing the measuring-rod, in just the same way as if all three were at rest.

b) By means of stationary clocks set up in the stationary system and synchronizing in accordance with § 1, the observer ascertains at what points of the stationary system the two ends of the rod to be measured are located at a definite time. The distance between these two points, measured by the measuring-rod already employed, which in this case is at rest, is also a length which may be designated "the length of the rod."

In accordance with the principle of relativity the length to be discovered by the operation a) – we will call it "the length of the rod in the moving system"– must be equal to the length l of the stationary rod.

The length to be discovered by the operation b) we will call "the length of the (moving) rod in the stationary system." This we shall determine on the basis of our two principles, and we shall find that it differs from l.

Current kinematics tacitly assumes that the lengths determined by these two operations are precisely equal, or in other words, that a moving rigid body at the epoch t may in geometrical respects be perfectly represented by *the same* body *at rest* in a definite position.

We imagine further that at the two ends A and B of the rod, clocks are placed which synchronize with the clocks of the stationary system, that is to say that their indications correspond at any instant to the "time of the stationary system" at the places where they happen to be. These clocks are therefore "synchronous in the stationary system."

We imagine further that with each clock there is a moving observer, and that these observers apply to both clocks the criterion established in § 1 for the synchronization of two clocks. Let a ray of light depart from A at the time[6] t_A, let it be reflected at B at the time t_B , and reach A again at the time t'_A. Taking into consideration the principle of the constancy of the velocity of light we find that

$$t_B - t_A = \frac{r_{AB}}{c - v} \quad \text{and} \quad t'_A - t_B = \frac{r_{AB}}{c + v},$$

where r_{AB} denotes the length of the moving rod – measured in the stationary system. Observers moving with the moving rod would thus find that the two clocks were not synchronous, while observers in the stationary system would declare the clocks to be synchronous.

So we see that we cannot attach any *absolute* signification to the concept of simultaneity, but that two events which, viewed[7] from a system of coordinates, are simultaneous, can no longer be looked upon as simultaneous events when envisaged from a system which is in motion relatively to that system.

§ 3. Theory of the Transformation of Coordinates and Times from a Stationary System to another System in Uniform Motion of Translation Relatively to the Former

Let us in "stationary" space take two systems of coordinates, i.e. two systems, each of three rigid material lines, perpendicular to one another, and issuing from a point. Let the axes of X of the two systems coincide, and their axes of Y and Z respectively be parallel. Let each system be provided with a rigid measuring-rod and a number of clocks, and let the two measuring-rods, and likewise all the clocks of the two systems, be in all respects alike.

Now to the origin of one of the two systems (k) let a constant velocity v be

---

[6]"Time" here denotes "time of the stationary system " and also "position of hands of the moving clock situated at the place under discussion."

[7]EDITOR'S NOTE: It should be stressed that in the theory of relativity "viewed" does *not* mean "seen." Einstein carefully described the procedure of determining which events are simultaneous. The relativistic effects have nothing to do with what we see; moreover we see only past events (Minkowski would say we see the past light cone).

imparted in the direction of the increasing x of the other stationary system (K), and let this velocity be communicated to the axes of the coordinates, the relevant measuring-rod, and the clocks. To any time of the stationary system K there then will correspond a definite position of the axes of the moving system, and from reasons of symmetry we are entitled to assume that the motion of k may be such that the axes of the moving system are at the time t (this "t" always denotes a time of the stationary system) parallel to the axes of the stationary system.

We now imagine space to be measured from the stationary system K by means of the stationary measuring-rod, and also from the moving system k by means of the measuring-rod moving with it; and that we thus obtain the coordinates x, y, z, and ξ, η, ζ, respectively. Further, let the time t of the stationary system be determined for all points thereof at which there are clocks by means of light signals in the manner indicated in § 1; similarly let the time τ of the moving system be determined for all points of the moving system at which there are clocks at rest relatively to that system by applying the method, given in § 1, of light signals between the points at which the latter clocks are located.

To any system of values x, y, z, t, which completely defines the place and time of an event in the stationary system, there belongs a system of values ξ, η, ζ, τ, determining that event relatively to the system k, and our task is now to find the system of equations connecting these quantities.

In the first place it is clear that the equations must be *linear* on account of the properties of homogeneity which we attribute to space and time.

If we place $x' = x - vt$, it is clear that a point at rest in the system k must have a system of values $x'y, z$, independent of time. We first define τ as a function of x', y, z, and t. To do this we have to express in equations that τ is nothing else than the summary of the data of clocks at rest in system k, which have been synchronized according to the rule given in § 1.

From the origin of system k let a ray be emitted at the time τ_0 along the X-axis to x', and at the time τ_1 be reflected thence to the origin of the coordinates, arriving there at the time τ_2; we then must have $\frac{1}{2}(\tau_0 + \tau_2) = \tau_1$, or, by inserting the arguments of the function τ and applying the principle of the constancy of the velocity of light in the stationary system:

$$\frac{1}{2}\left[\tau(0,0,0,t) + \tau\left(0,0,0,t + \frac{x'}{c-v} + \frac{x'}{c+v}\right)\right] = \tau\left(x',0,0,t + \frac{x'}{c-v}\right).$$

Hence, if x' be chosen infinitesimally small,

$$\frac{1}{2}\left(\frac{1}{c-v} + \frac{1}{c+v}\right)\frac{\partial\tau}{\partial t} = \frac{\partial\tau}{\partial x'} + \frac{1}{c-v}\frac{\partial\tau}{\partial t},$$

or

$$\frac{\partial\tau}{\partial x'} + \frac{v}{c^2-v^2}\frac{\partial\tau}{\partial t} = 0.$$

It is to be noted that instead of the origin of the coordinates we might have chosen any other point for the point of origin of the ray, and the equation just obtained is therefore valid for all values of x', y, z.

An analogous consideration – applied to the axes of Y and Z – it being borne in mind that light is always propagated along these axes, when viewed from the stationary system, with the velocity $\sqrt{(c^2 - v^2)}$, gives us

$$\frac{\partial \tau}{\partial y} = 0, \quad \frac{\partial \tau}{\partial z} = 0.$$

Since τ is a *linear* function, it follows from these equations that

$$\tau = a\left(t - \frac{v}{c^2 - v^2} x'\right)$$

where a is a function $\phi(v)$ at present unknown, and where for brevity it is assumed that at the origin of k, $\tau = 0$, when $t = 0$.

With the help of this result we easily determine the quantities ξ, η, ζ by expressing in equations that light (as required by the principle of the constancy of the velocity of light, in combination with the principle of relativity) is also propagated with velocity c when measured in the moving system. For a ray of light emitted at the time $\tau = 0$ in the direction of the increasing ξ

$$\xi = c\tau \quad \text{or} \quad \xi = ac\left(t - \frac{v}{c^2 - v^2} x'\right).$$

But the ray moves relatively to the initial point of k, when measured in the stationary system, with the velocity $c - v$, so that

$$\frac{x'}{c - v} = t.$$

If we insert this value of t in the equation for ξ we obtain

$$\xi = a\frac{c^2}{c^2 - v^2} x'.$$

In an analogous manner we find, by considering rays moving along the two other axes, that

$$\eta = c\tau = ac\left(t - \frac{v}{c^2 - v^2} x'\right)$$

when

$$\frac{y}{\sqrt{c^2 - v^2}} = t, \quad x' = 0.$$

Thus

$$\eta = a\frac{c}{\sqrt{c^2 - v^2}} y \quad \text{and} \quad \zeta = a\frac{c}{\sqrt{c^2 - v^2}} z.$$

Substituting for x' its value, we obtain

$$\tau = \phi(v)\beta(t - vx/c^2),$$
$$\xi = \phi(v)\beta(x - vt),$$
$$\eta = \phi(v)y,$$
$$\zeta = \phi(v)z,$$

where

$$\beta = \frac{1}{\sqrt{1 - \frac{v^2}{c^2}}}$$

and ϕ is an as yet unknown function of v. If no assumption whatever be made as to the initial position of the moving system and as to the zero point of τ, an additive constant is to be placed on the right side of each of these equations.

We now have to prove that any ray of light, measured in the moving system, is propagated with the velocity c, if, as we have assumed, this is the case in the stationary system; for we have not as yet furnished the proof that the principle of the constancy of the velocity of light is compatible with the principle of relativity.

At the time $t = \tau = 0$, when the origin of the coordinates is common to the two systems, let a spherical wave be emitted therefrom, and be propagated with the velocity c in system K. If (x, y, z) be a point just attained by this wave, then

$$x^2 + y^2 + z^2 = c^2 t^2.$$

Transforming this equation with the aid of our equations of transformation we obtain after a simple calculation

$$\xi^2 + \eta^2 + \zeta^2 = c^2 \tau^2.$$

The wave under consideration is therefore no less a spherical wave with velocity of propagation c when viewed in the moving system. This shows that our two fundamental principles are compatible.[8]

In the equations of transformation which have been developed there enters an unknown function ϕ of v which we will now determine.

For this purpose we introduce a third system of coordinates K' which relatively to the system k is in a state of parallel translatory motion parallel to k's axis of Ξ, such that its (K''s) origin of coordinates moves with velocity $-v$ along the axis of Ξ.[9] At the time $t = 0$ let all three origins coincide, and when $t = x = y = z = 0$

---

[8]The equations of the Lorentz transformation may be more simply deduced directly from the condition that in virtue of those equations the relation $x^2 + y^2 + z^2 = c^2 t^2$ shall have as its consequence the second relation $\xi^2 + \eta^2 + \zeta^2 = c^2 \tau^2$.

[9]EDITOR'S NOTE: The 1923 English translation of this sentence contains an error (it should be v, not $-v$) [1]:

> For this purpose we introduce a third system of coordinates K' which relatively to the system k is in a state of parallel translatory motion parallel to the axis of X, such that the origin of coordinates of system k moves with velocity $-v$ on the axis of X.

By definition (p. 83) the system k moves with velocity v (not $-v$) along the X axis of the "stationary" system K. The corresponding German sentence says that it is the system K' (not k) that moves with velocity $-v$ along k's axis Ξ (not along K's axis X) [2]:

> Wir führen zu diesem Zwecke noch ein drittes Koordinatensystem K' ein, welches relativ zum System k derart in Paralleltranslationsbewegung parallel zur Ξ-Achse begriffen sei, daß sich dessen Koordinatenursprung mit der Geschwindigkeit $-v$ auf der Ξ-Achse bewege.

This means that K and K' are at rest with respect to each other as seen from Einstein's text below. The translation of the sentence included here closely follows the original German sentence.

let the time t' of the system K' be zero. We call the coordinates, measured in the system K', x', y', z', and by a twofold application of our equations of transformation we obtain

$$
\begin{aligned}
t' &= \phi(-v)\beta(-v)(\tau + v\xi/c^2) &&= \phi(v)\phi(-v)t, \\
x' &= \phi(-v)\beta(-v)(\xi + v\tau) &&= \phi(v)\phi(-v)x, \\
y' &= \phi(-v)\eta &&= \phi(v)\phi(-v)y, \\
z' &= \phi(-v)\zeta &&= \phi(v)\phi(-v)z.
\end{aligned}
$$

Since the relations between x', y', z' and x, y, z do not contain the time t, the systems K and K' are at rest with respect to one another, and it is clear that the transformation from K to K' must be the identical transformation. Thus

$$
\phi(v)\phi(-v) = 1.
$$

We now inquire into the signification of $\phi(v)$. We give our attention to that part of the axis of Y of system k which lies between $\xi = 0, \eta = 0, \zeta = 0$ and $\xi = 0, \eta = l$, $\zeta = 0$. This part of the axis of Y is a rod moving perpendicularly to its axis with velocity v relatively to system K. Its ends possess in K the coordinates

$$
x_1 = vt, \quad y_1 = \frac{l}{\phi(v)}, \quad z_1 = 0
$$

and

$$
x_2 = vt, \quad y_2 = 0, \quad z_2 = 0.
$$

The length of the rod measured in K is therefore $l/\phi(v)$; and this gives us the meaning of the function $\phi(v)$. From reasons of symmetry it is now evident that the length of a given rod moving perpendicularly to its axis, measured in the stationary system, must depend only on the velocity and not on the direction and the sense of the motion. The length of the moving rod measured in the stationary system does not change, therefore, if v and $-v$ are interchanged. Hence it follows that

$$
\frac{l}{\phi(v)} = \frac{l}{\phi(-v)},
$$

or

$$
\phi(v) = \phi(-v).
$$

I would like to thank Jan Pilotti for drawing my attention to this error.

[1] H. A. Lorentz, A. Einstein, H. Minkowski and H. Weyl, *The Principle of Relativity: A Collection of Original Memoirs on the Special and General Theory of Relativity*. With Notes by A. Sommerfeld. Translated by W. Perrett and G. B. Jeffery (Methuen and Company Ltd., 1923; reprinted by Dover Puplications Inc., 1952), p. 47.

[2] H. A. Lorentz, A. Einstein, H. Minkowski, *Das Relativitätsprinzip: Eine Sammlung von Abhandlungen*. Mit einem Beitrag von H. Weyl und Anmerkungen von A. Sommerfeld. Vorwort von O. Blumenthal, Fünfte Auflage (Springer Fachmedien Wiesbaden Gmbh, 1923), p. 34.

It follows from this relation and the one previously found that $\phi(v) = 1$, so that the transformation equations which have been found become

$$\tau = \beta)t - vx/c^2),$$
$$\xi = \beta(x - vt),$$
$$\eta = y,$$
$$\zeta = z,$$

where

$$\beta = \frac{1}{\sqrt{1 - \frac{v^2}{c^2}}}.$$

§ 4. Physical Meaning of the Equations Obtained in Respect to Moving Rigid Bodies and Moving Clocks

We envisage a rigid sphere[10] of radius R, at rest relatively to the moving system k, and with its centre at the origin of coordinates of k. The equation of the surface of this sphere moving relatively to the system K with velocity v is

$$\xi^2 + \eta^2 + \xi^2 = R^2$$

The equation of this surface expressed in x, y, z at the time $t = 0$ is

$$\frac{x^2}{\left(\sqrt{1 - \frac{v^2}{c^2}}\right)^2} + y^2 + z^2 = R^2.$$

A rigid body which, measured in a state of rest, has the form of a sphere, therefore has in a state of motion – viewed from the stationary system – the form of an ellipsoid of revolution with the axes

$$R\sqrt{1 - \frac{v^2}{c^2}}, \; R, \; R.$$

Thus, whereas the Y and Z dimensions of the sphere (and therefore of every rigid body of no matter what form) do not appear modified by the motion, the X dimension appears shortened in the ratio $1 : \sqrt{1 - v^2/c^2}$, i.e. the greater the value of v, the greater the shortening. For $v = c$ all moving objects – viewed from the 'stationary" system – shrivel up into plain figures. For velocities greater than that of light our deliberations become meaningless; we shall, however, find in what follows, that the velocity of light in our theory plays the part, physically, of an infinitely great velocity.

It is clear that the same results hold good of bodies at rest in the "stationary" system, viewed from a system in uniform motion.

Further, we imagine one of the clocks which are qualified to mark the time t when at rest relatively to the stationary system, and the time τ when at rest relatively

---

[10]That is, a body possessing spherical form when examined at rest.

to the moving system, to be located at the origin of the coordinates of k , and so adjusted that it marks the time τ. What is the rate of this clock, when viewed from the stationary system?

Between the quantities x, t, and τ, which refer to the position of the clock, we have, evidently, $x = vt$ and

$$\tau = \frac{1}{\sqrt{1 - \frac{v^2}{c^2}}} \left(t - \frac{vx}{c^2} \right).$$

Therefore,

$$\tau = t\sqrt{1 - \frac{v^2}{c^2}} = t - \left(\sqrt{1 - \frac{v^2}{c^2}} \right) t,$$

whence it follows that the time marked by the clock (viewed in the stationary system) is slow by $1 - \sqrt{1 - v^2/c^2}$ seconds per second, or – neglecting magnitudes of fourth and higher order – by $\frac{1}{2} v^2/c^2$.

From this there ensues the following peculiar consequence. If at the points A and B of K there are stationary clocks which, viewed in the stationary system, are synchronous, and if the clock at A is moved with the velocity v along the line AB to B, then on its arrival at B the two clocks no longer synchronized, but the clock moved from A to B lags behind the other which has remained at B by $\frac{1}{2}t\,(v/c)^2$ seconds (up to magnitudes of fourth and higher order), where t is the clock's time to travel from A to B.

It is at once apparent that this result still holds good if the clock moves from A to B in any polygonal line, and also when the points A and B coincide.

If we assume that the result proved for a polygonal line is also valid for a continuously curved line, we arrive at this result: If one of two synchronous clocks at A is moved in a closed curve with constant speed until it returns to A, the journey lasting t seconds, then by the clock which has remained at rest the traveled clock on its arrival at A will be $\frac{1}{2}t(v/c)^2$ seconds slow. Thence we conclude that a balance-clock[11] at the equator must go more slowly, by a very small amount, than a precisely similar clock situated at one of the poles under otherwise identical conditions.

§ 5. The Composition of Velocities

In the system k moving along the axis of X of the system K with velocity v, let a point move in accordance with the equations

$$\xi = w_\xi \tau,$$
$$\eta = w_\eta \tau,$$
$$\zeta = 0,$$

where w_ξ and w_η denote constants.

---

[11] Not a pendulum-clock, which is physically a system to which the Earth belongs. This case had to be excluded.

Required: the motion of the point relatively to the system K. If with the help of the equations of transformation developed in § 3 we introduce the quantities x, y, z, t into the equations of motion of the point, we obtain

$$x = \frac{\omega_\xi + v}{1 + vw_\xi/c^2}t,$$

$$y = \frac{\sqrt{1 - v^2/c^2}}{1 + vw_\xi/c^2}w_\eta t,$$

$$z = 0$$

Thus the law of the parallelogram of velocities is valid according to our theory only to a first approximation. We set

$$V^2 = \left(\frac{dx}{dt}\right)^2 + \left(\frac{dy}{dt}\right)^2,$$

$$w^2 = w_\xi^2 + w_\eta^2,$$

$$\alpha = \tan^{-1}\frac{w_y}{w_x},$$

α is then to be looked upon as the angle between the velocities v and w. After a simple calculation we obtain

$$V = \frac{\sqrt{(v^2 + w^2 + 2vw\cos\alpha) - \left(\frac{vw\sin\alpha}{c}\right)^2}}{1 + \frac{vw\cos\alpha}{c^2}}.$$

It is worthy of remark that v and w enter into the expression for the resultant velocity in a symmetrical manner. If w also has the direction of the axis of X, we get

$$V = \frac{v + w}{1 + \frac{vw}{c^2}}.$$

It follows from this equation that from a composition of two velocities which are less than c, there always results a velocity less than c. For if we set $v = c - \kappa$, $w = c - \lambda$, κ and λ being positive and less than c, then

$$V = c\frac{2c - \kappa - \lambda}{2c - \kappa - \lambda + \kappa\lambda/c} < c.$$

It follows, further, that the velocity of light c cannot be altered by composition with a velocity less than that of light. For this case we obtain

$$V = \frac{c + w}{1 + w/c} = c$$

We might also have obtained the formula for V, for the case when v and w have the same direction, by compounding two transformations in accordance with § 3 . If in addition to the systems K and k figuring in § 3 we introduce still another system of coordinates k' moving parallel to k, its initial point moving on the axis of X with the velocity w, we obtain equations between the quantities x, y, z, t and the corresponding quantities of k', which differ from the equations found in § 3 only in that the place of "v" is taken by the quantity

$$\frac{v+w}{1+vw/c^2},$$

from which we see that such parallel transformations – necessarily – form a group.

We have now deduced the requisite laws of the theory of kinematics corresponding to our two principles, and we proceed to show their application to electrodynamics.

II. Electrodynamical Part

§ 6. Transformation of the Maxwell-Hertz Equations for Empty Space. On the Nature of the Electromotive Forces Occurring in a Magnetic Field During Motion

Let the Maxwell-Hertz equations for empty space hold good for the stationary system K, so that we have

$$\frac{1}{c}\frac{\partial X}{\partial t} = \frac{\partial N}{\partial y} - \frac{\partial M}{\partial z}, \quad \frac{1}{c}\frac{\partial L}{\partial t} = \frac{\partial Y}{\partial z} - \frac{\partial Z}{\partial y},$$

$$\frac{1}{c}\frac{\partial Y}{\partial t} = \frac{\partial L}{\partial z} - \frac{\partial N}{\partial x}, \quad \frac{1}{c}\frac{\partial M}{\partial t} = \frac{\partial Z}{\partial x} - \frac{\partial X}{\partial z},$$

$$\frac{1}{c}\frac{\partial Z}{\partial t} = \frac{\partial M}{\partial x} - \frac{\partial L}{\partial y}, \quad \frac{1}{c}\frac{\partial N}{\partial t} = \frac{\partial X}{\partial y} - \frac{\partial Y}{\partial x},$$

where (X, Y, Z) denotes the vector of the electric force, and (L, M, N) that of the magnetic force.

If we apply to these equations the transformation developed in § 3 , by referring the electromagnetic processes to the system of coordinates there introduced, moving with the velocity v, we obtain the equations

$$\frac{1}{c}\frac{\partial X}{\partial \tau} = \frac{\partial}{\partial \eta}\left\{\beta\left(N - \frac{v}{c}Y\right)\right\} - \frac{\partial}{\partial \zeta}\left\{\beta\left(M + \frac{v}{c}Z\right)\right\},$$

$$\frac{1}{c}\frac{\partial}{\partial \tau}\left\{\beta\left(Y - \frac{v}{c}N\right)\right\} = \frac{\partial L}{\partial \xi} - \frac{\partial}{\partial \zeta}\left\{\beta\left(N - \frac{v}{c}\right)Y\right\},$$

$$\frac{1}{c}\frac{\partial}{\partial \tau}\left\{\beta\left(Z + \frac{v}{c}M\right)\right\} = \frac{\partial}{\partial \xi}\left\{\beta\left(M + \frac{v}{c}Z\right)\right\} - \frac{\partial L}{\partial \eta},$$

$$-\frac{1}{c}\frac{\partial L}{\partial \tau} = \frac{\partial}{\partial \zeta}\left\{\beta\left(Y - \frac{v}{c}N\right)\right\} - \frac{\partial}{\partial \eta}\left\{\beta\left(Z + \frac{v}{c}M\right)\right\},$$

$$\frac{1}{c}\frac{\partial}{\partial \tau}\left\{\beta\left(M + \frac{v}{c}Z\right)\right\} = \frac{\partial}{\partial \xi}\left\{\beta\left(Z + \frac{v}{c}M\right)\right\} - \frac{\partial X}{\partial \zeta},$$

$$\frac{1}{c}\frac{\partial}{\partial \tau}\left\{\beta\left(N - \frac{v}{c}Y\right)\right\} = \frac{\partial X}{\partial \eta} - \frac{\partial}{\partial \xi}\left\{\beta\left(Y - \frac{v}{c}N\right)\right\},$$

where

$$\beta = \frac{1}{\sqrt{1 - v^2/c^2}}.$$

Now the principle of relativity requires that if the Maxwell-Hertz equations for empty space hold good in system K, they also hold good in system k ; that is to say that the vectors of the electric and the magnetic force – (X', Y', Z') and (L', M', N') – of the moving system k , which are defined by their ponderomotive effects on electric or magnetic masses respectively, satisfy the following equations : –

$$\frac{1}{c}\frac{\partial X'}{\partial \tau} = \frac{\partial N'}{\partial \eta} - \frac{\partial M'}{\partial \zeta}, \qquad \frac{1}{c}\frac{\partial L'}{\partial \tau} = \frac{\partial Y'}{\partial \zeta} - \frac{\partial Z'}{\partial \eta},$$

$$\frac{1}{c}\frac{\partial Y'}{\partial \tau} = \frac{\partial L'}{\partial \zeta} - \frac{\partial N'}{\partial \xi}, \qquad \frac{1}{c}\frac{\partial M'}{\partial \tau} = \frac{\partial Z'}{\partial \xi} - \frac{\partial X'}{\partial \zeta},$$

$$\frac{1}{c}\frac{\partial Z'}{\partial \tau} = \frac{\partial M'}{\partial \xi} - \frac{\partial L'}{\partial \eta}, \qquad \frac{1}{c}\frac{\partial N'}{\partial \tau} = \frac{\partial X'}{\partial \eta} - \frac{\partial Y'}{\partial \xi}.$$

Evidently the two systems of equations found for system k must express exactly the same thing, since both systems of equations are equivalent to the Maxwell-Hertz equations for system K. Since, further, the equations of the two systems agree, with the exception of the symbols for the vectors, it follows that the functions occurring in the systems of equations at corresponding places must agree, with the exception of a factor $\psi(v)$, which is common for all functions of the one system of equations, and is independent of ξ, η, ζ and τ but depends upon v. Thus we have the relations

$$X' = \psi(v)X, \qquad\qquad L' = \psi(v)L,$$
$$Y' = \psi(v)\beta\left(Y - \tfrac{v}{c}N\right), \quad M' = \psi(v)\beta\left(M + \tfrac{v}{c}Z\right),$$
$$Z' = \psi(v)\beta\left(Z + \tfrac{v}{c}M\right), \quad N' = \psi(v)\beta\left(N - \tfrac{v}{c}Y\right).$$

If we now form the reciprocal of this system of equations, firstly by solving the equations just obtained, and secondly by applying the equations to the inverse transformation (from k to K), which is characterized by the velocity - v, it follows, when we consider that the two systems of equations thus obtained must be identical, that

$$\psi(v) \cdot \psi(-v) = 1.$$

Further, from reasons of symmetry[12]

$$\psi(v) = \psi(-v);$$

and therefore

$$\psi(v) = 1$$

---

[12] If, for example, $X = Y = Z = L = M = 0$, and $N \neq 0$, then from reasons of symmetry it is clear that when v changes sign without changing its numerical value, Y' must also change sign without changing its numerical value.

and our equations assume the form

$$X' = X, \qquad\qquad L' = L,$$
$$Y' = \beta\left(Y - \tfrac{v}{c}N\right), \quad M' = \beta\left(M + \tfrac{v}{c}Z\right),$$
$$Z' = \beta\left(Z + \tfrac{v}{c}M\right), \quad N' = \beta\left(N - \tfrac{v}{c}Y\right).$$

As to the interpretation of these equations we make the following remarks: Let a point charge of electricity have the magnitude "one" when measured in the stationary system K, i.e. let it when at rest in the stationary system exert a force of one dyne upon an equal quantity of electricity at a distance of one cm. By the principle of relativity this electric charge is also of the magnitude "one" when measured in the moving system. If this quantity of electricity is at rest relatively to the stationary system, then by definition the vector (X, Y, Z) is equal to the force acting upon it. If the quantity of electricity is at rest relatively to the moving system (at least at the relevant instant), then the force acting upon it, measured in the moving system, is equal to the vector (X', Y', Z'). Consequently the first three equations above allow themselves to be clothed in words in the two following ways:

1. If a unit electric point charge is in motion in an electromagnetic field, there acts upon it, in addition to the electric force, an "electromotive force" which, if we neglect the terms multiplied by the second and higher powers of v/c, is equal to the vector-product of the velocity of the charge and the magnetic force, divided by the velocity of light. (Old manner of expression.)

2. If a unit electric point charge is in motion in an electromagnetic field, the force acting upon it is equal to the electric force which is present at the locality of the charge, and which we ascertain by transformation of the field to a system of coordinates at rest relatively to the electrical charge. (New manner of expression.)

The analogy holds with "magnetomotive forces." We see that electromotive force plays in the developed theory merely the part of an auxiliary concept, which owes its introduction to the circumstance that electric and magnetic forces do not exist independently of the state of motion of the system of coordinates.

Furthermore it is clear that the asymmetry mentioned in the introduction as arising when we consider the currents produced by the relative motion of a magnet and a conductor, now disappears. Moreover, questions as to the "seat" of electrodynamic electromotive forces (unipolar machines) now have no point.

§ 7. Theory of Doppler's Principle and of Aberration

In the system K, very far from the origin of coordinates, let there be a source of electrodynamic waves, which in a part of space containing the origin of coordinates may be represented to a sufficient degree of approximation by the equations

$$X = X_0 \sin \Phi, \quad L = L_0 \sin \Phi,$$
$$Y = Y_0 \sin \Phi, \quad M = M_0 \sin \Phi,$$
$$Z = Z_0 \sin \Phi, \quad N = N_0 \sin \Phi,$$

where

$$\Phi = \omega \left\{ t - \frac{1}{c}(lx + my + nz) \right\}.$$

Here (X_0, Y_0, Z_0) and (L_0, M_0, N_0) are the vectors defining the amplitude of the wave-train, and l, m,n the direction-cosines of the wave-normals. We wish to know the constitution of these waves, when they are examined by an observer at rest in the moving system k.

Applying the equations of transformation found in § 6 for electric and magnetic forces, and those found in § 3 for the coordinates and the time, we obtain directly

$$X' = X_0 \sin \Phi', \qquad\qquad L' = L_0 \sin \Phi'$$
$$Y' = \beta(Y_0 - vN_0/c) \sin \Phi', \qquad M' = \beta(M_0 + vZ_0/c) \sin \Phi',$$
$$Z' = \beta(Z_o + vM_o/c) \sin \Phi', \qquad N' = \beta(N_0 - vY_0/c) \sin \Phi',$$
$$\Phi' = \omega' \left\{ \tau - \tfrac{1}{c}(l'\xi + m'\eta + n'\zeta) \right\},$$

where

$$\omega' = \omega\beta(1 - lv/c),$$
$$l' = \frac{l - v/c}{1 - lv/c},$$
$$m' = \frac{m}{\beta(1 - lv/c)},$$
$$n' = \frac{n}{\beta(1 - lv/c)}.$$

From the equation for ω' it follows that if an observer is moving with velocity v relatively to an infinitely distant source of light of frequency ν, in such a way that the connecting line "source–observer" makes the angle ϕ with the velocity of the observer referred to a system of coordinates which is at rest relatively to the source of light, the frequency ν' of the light perceived by the observer is given by the equation

$$\nu' = \nu \frac{1 - v/c \cdot \cos\phi}{\sqrt{1 - v^2/c^2}}$$

This is Doppler's principle for any velocities whatever. When $\phi = 0$ the equation assumes the perspicuous form

$$\nu' = \nu \sqrt{\frac{1 - v/c}{1 + v/c}}.$$

We see that, in contrast with the customary view, when $v = -\infty$, $\nu' = \infty$.

If we call the angle between the wave-normal (direction of the ray) in the moving system and the connecting line "source—observer " ϕ', the equation for l' assumes the form

$$\cos\phi' = \frac{\cos\phi - v/c}{1 - v/c \cdot \cos\phi}.$$

This equation expresses the law of aberration in its most general form. If $\phi = \pi/2$, the equation becomes simply

$$\cos \phi' = -\frac{v}{c}.$$

We still have to find the amplitude of the waves, as it appears in the moving system. If we call the amplitude of the electric or magnetic force A or A' respectively, accordingly as it is measured in the stationary system or in the moving system, we obtain

$$A'^2 = A^2 \frac{(1 - v/c \cdot \cos \phi)^2}{1 - v^2/c^2}$$

which equation, if $\phi = 0$, simplifies into

$$A'^2 = A^2 \frac{1 - v/c}{1 + v/c}.$$

It follows from these results that to an observer approaching a source of light with the velocity c, this source of light must appear of infinite intensity.

§ 8. Transformation of the Energy of Light Rays. Theory of the Pressure of Radiation Exerted on Perfect Reflectors

Since $A^2/8\pi$ equals the energy of light per unit of volume, we have to regard $A'^2/8\pi$, by the principle of relativity, as the energy of light in the moving system. Thus A'^2/A^2 would be the ratio of the "measured in motion" to the "measured at rest" energy of a given light complex, if the volume of a light complex were the same, whether measured in K or in k. But this is not the case. If l, m, n are the direction-cosines of the wave-normals of the light in the stationary system, no energy passes through the surface elements of a spherical surface moving with the velocity of light:

$$(x - lct)^2 + (y - mct)^2 + (z - nct)^2 = R^2.$$

We may therefore say that this surface permanently encloses the same light complex. We inquire as to the quantity of energy enclosed by this surface, viewed in system k, that is, as to the energy of the light complex relatively to the system k.

The spherical surface—viewed in the moving system – is an ellipsoidal surface, the equation for which, at the time $\tau = 0$, is

$$(\beta\xi - l\beta\xi v/c)^2 + (\eta - m\beta\xi v/c)^2 + (\zeta - n\beta\xi v/c)^2 = R^2.$$

If S is the volume of the sphere, and S' that of this ellipsoid, then by a simple calculation

$$\frac{S'}{S} = \frac{\sqrt{1 - v^2/c^2}}{1 - v/c \cdot \cos \phi}$$

Thus, if we call the light energy enclosed by this surface E when it is measured in the stationary system, and E' when measured in the moving system, we obtain

$$\frac{E'}{E} = \frac{A'^2 S'}{A^2 S} = \frac{1 - v/c \cdot \cos \phi}{\sqrt{1 - v^2/c^2}}$$

and this formula, when $\phi = 0$, simplifies into

$$\frac{E'}{E} = \sqrt{\frac{1 - v/c}{1 + v/c}}.$$

It is remarkable that the energy and the frequency of a light complex vary with the state of motion of the observer in accordance with the same law.

Now let the coordinate plane $\xi = 0$ be a perfectly reflecting surface, at which the plane waves considered in § 7 are reflected. We seek for the pressure of light exerted on the reflecting surface, and for the direction, frequency, and intensity of the light after reflection.

Let the incidental light be defined by the quantities A, $\cos \phi$, ν (referred to system K). Viewed from k the corresponding quantities are

$$A' = A \frac{1 - v/c \cdot \cos \phi}{\sqrt{1 - v^2/c^2}},$$

$$\cos \phi' = \frac{\cos \phi - v/c}{1 - v/c \cdot \cos \phi},$$

$$\nu' = \nu \frac{1 - v/c \cdot \cos \phi}{\sqrt{1 - v^2/c^2}}.$$

For the reflected light, referring the process to system k, we obtain

$$A'' = A'$$
$$\cos \phi'' = -\cos \phi'$$
$$\nu'' = nu'$$

Finally, by transforming back to the stationary system K, we obtain for the reflected light

$$A''' = A'' \frac{1 + v/c \cdot \cos \phi''}{\sqrt{1 - v^2/c^2}} = A \frac{1 - 2v/c \cdot \cos \phi + v^2/c^2}{1 - v^2/c^2}$$

$$\cos \phi''' = \frac{\cos \phi'' + v/c}{1 + v/c \cdot \cos \phi''} = -\frac{(1 + v^2/c^2) \cos \phi - 2v/c}{1 - 2v/c \cdot \cos \phi + v^2/c^2}$$

$$\nu''' = \nu'' \frac{1 + v/c \cdot \cos \phi''}{\sqrt{1 - v^2/c^2}} = \nu \frac{1 - 2v/c \cdot \cos \phi + v^2/c^2}{(1 - v/c)^2}.$$

The energy (measured in the stationary system) which is incident upon unit area of the mirror in unit time is evidently $A^2(c \cos \phi - v)/8\pi$. The energy leaving the unit of surface of the mirror in the unit of time is $A'''^2(-c \cos \phi''' + v)/8\pi$ The difference of these two expressions is, by the principle of energy, the work done by the pressure of light in the unit of time. If we set down this work as equal to the product $P \cdot v$ where P is the pressure of light, we obtain

$$P = 2 \frac{A^2}{8\pi} \frac{(\cos \phi - v/c)^2}{1 - v^2/c^2}.$$

In agreement with experiment and with other theories, we obtain to a first approximation

$$P = 2 \frac{A^2}{8\pi} \cos^2 \phi.$$

All problems in the optics of moving bodies can be solved by the method here employed. What is essential is, that the electric and magnetic force of the light which is influenced by a moving body, be transformed into a system of coordinates at rest relatively to the body. By this means all problems in the optics of moving bodies will be reduced to a series of problems in the optics of stationary bodies.

§ 9. Transformation of the Maxwell-Hertz Equations when Convection-Currents are Taken into Account

We start from the equations

$$\frac{1}{c}\left\{u_x\rho + \frac{\partial X}{\partial t}\right\} = \frac{\partial N}{\partial y} - \frac{\partial M}{\partial z}, \quad \frac{1}{c}\frac{\partial L}{\partial t} = \frac{\partial Y}{\partial z} - \frac{\partial Z}{\partial y},$$

$$\frac{1}{c}\left\{u_y\rho + \frac{\partial Y}{\partial t}\right\} = \frac{\partial L}{\partial z} - \frac{\partial N}{\partial x}, \quad \frac{1}{c}\frac{\partial M}{\partial t} = \frac{\partial Z}{\partial x} - \frac{\partial X}{\partial z},$$

$$\frac{1}{c}\left\{u_z\rho + \frac{\partial Z}{\partial t}\right\} = \frac{\partial M}{\partial x} - \frac{\partial L}{\partial y}, \quad \frac{1}{c}\frac{\partial N}{\partial t} = \frac{\partial X}{\partial y} - \frac{\partial Y}{\partial x},$$

where

$$\rho = \frac{\partial X}{\partial x} + \frac{\partial Y}{\partial y} + \frac{\partial Z}{\partial z}$$

denotes 4π times the density of electricity, and (u_x, u_y, u_z) the velocity-vector of the charge. If we imagine the electric charges to be invariably coupled to small rigid bodies (ions, electrons), these equations are the electromagnetic basis of the Lorentzian electrodynamics and optics of moving bodies.

Let these equations be valid in the system K, and transform them, with the assistance of the equations of transformation given in §§ 3 and 6, to the system k. We then obtain the equations

$$\frac{1}{c}\left\{u_\xi\rho' + \frac{\partial X'}{\partial \tau}\right\} = \frac{\partial N'}{\partial \eta} - \frac{\partial M'}{\partial \zeta}, \quad \frac{1}{c}\frac{\partial L'}{\partial \tau} = \frac{\partial Y'}{\partial \zeta} - \frac{\partial Z'}{\partial \eta},$$

$$\frac{1}{c}\left\{u_\eta\rho' + \frac{\partial Y'}{\partial \tau}\right\} = \frac{\partial L'}{\partial \zeta} - \frac{\partial N'}{\partial \xi}, \quad \frac{1}{c}\frac{\partial M'}{\partial \tau} = \frac{\partial Z'}{\partial \xi} - \frac{\partial X'}{\partial \zeta},$$

$$\frac{1}{c}\left\{u_\zeta\rho' + \frac{\partial Z'}{\partial \tau}\right\} = \frac{\partial M'}{\partial \xi} - \frac{\partial L'}{\partial \eta}, \quad \frac{1}{c}\frac{\partial N'}{\partial \tau} = \frac{\partial X'}{\partial \eta} - \frac{\partial Y'}{\partial \xi'},$$

where

$$u_\xi = \frac{u_x - v}{1 - u_x v/c^2},$$

$$u_\eta = \frac{u_y}{\beta(1 - u_x v/c^2)},$$

$$u_\zeta = \frac{u_z}{\beta(1 - u_x v/c^2)},$$

and

$$\rho' = \frac{\partial X'}{\partial \xi} + \frac{\partial Y'}{\partial \eta} + \frac{\partial Z'}{\partial \zeta} = \beta(1 - u_x v/c^2)\rho.$$

Since – as follows from the theorem of addition of velocities (§ 5)—the vector (u_ξ, u_η, u_ζ) is nothing else than the velocity of the electric charge, measured in the system k, we have the proof that, on the basis of our kinematical principles, the electrodynamic foundation of Lorentz's theory of the electrodynamics of moving bodies is in agreement with the principle of relativity.

In addition I may briefly remark that the following important law may easily be deduced from the developed equations: If an electrically charged body is in motion anywhere in space without altering its charge when regarded from a system of coordinates moving with the body, its charge also remains – when regarded from the "stationary" system K – constant.

§ 10. Dynamics of the Slowly Accelerated Electron

Let there be in motion in an electromagnetic field an electrically charged particle (in the sequel called an "electron"), for the law of motion of which we assume as follows:

If the electron is at rest at a given epoch, the motion of the electron ensues in the next instant of time according to the equations

$$m\frac{d^2 x}{dt^2} = \epsilon X,$$

$$m\frac{d^2 y}{dt^2} = \epsilon Y,$$

$$m\frac{d^2 z}{dt^2} = \epsilon Z,$$

where x, y, z denote the coordinates of the electron, and m the mass of the electron, as long as its motion is slow.

Now, secondly, let the velocity of the electron at a given epoch be v. We seek the law of motion of the electron in the immediately ensuing instants of time.

Without affecting the general character of our considerations, we may and will assume that the electron, at the moment when we give it our attention, is at the origin of the coordinates, and moves with the velocity v along the axis of X of the system K. It is then clear that at the given moment ($t = 0$) the electron is at rest relatively to a system of coordinates which is in parallel motion with velocity v along the axis of X.

From the above assumption, in combination with the principle of relativity, it is clear that in the immediately ensuing time (for small values of t) the electron, viewed from the system k, moves in accordance with the equations

$$m\frac{d^2\xi}{d\tau^2} = \epsilon X',$$

$$m\frac{d^2\eta}{d\tau^2} = \epsilon Y',$$

$$m\frac{d^2\zeta}{d\tau^2} = \epsilon Z',$$

in which the symbols $\xi, \eta, \zeta, \tau, X', Y', Z'$ refer to the system k. If, further, we decide that when $t = x = y = z = 0$ then $\tau = \xi = \eta = \zeta = 0$, the transformation equations of §§ 3 and 6 hold good, so that we have

$$\tau = \beta(t - vx/c^2),$$

$$\xi = \beta(x - vt), \qquad X' = X,$$

$$\eta = y, \qquad Y' = \beta(Y - vN/c),$$

$$\zeta = z, \qquad Z' = \beta(Z + vM/c)$$

With the help of these equations we transform the above equations of motion from system k to system K, and obtain

$$\left.\begin{array}{l} \dfrac{d^2x}{dt^2} = \dfrac{\epsilon}{m\beta^3}X \\[2ex] \dfrac{d^2y}{dt^2} = \dfrac{\epsilon}{m\beta}\left(Y - \dfrac{v}{c}N\right) \\[2ex] \dfrac{d^2z}{dt^2} = \dfrac{\epsilon}{m\beta}\left(Z + \dfrac{v}{c}M\right) \end{array}\right\}. \qquad (A)$$

Taking the ordinary point of view we now inquire as to the "longitudinal" and the "transverse" mass of the moving electron. We write the equations $A)$ in the form

$$m\beta^3\frac{d^2x}{dt^2} = \epsilon X = \epsilon X',$$

$$m\beta^2\frac{d^2y}{dt^2} = \epsilon\beta\left(Y - \frac{v}{c}N\right) = \epsilon Y',$$

$$m\beta^2\frac{d^2z}{dt^2} = \epsilon\beta\left(Z + \frac{v}{c}M\right) = \epsilon Z',$$

and remark firstly that $\epsilon X', \epsilon Y', \epsilon Z'$ are the components of the ponderomotive force acting upon the electron, and are so indeed as viewed in a system moving at the moment with the electron, with the same velocity as the electron. (This force might be measured, for example, by a spring balance at rest in the last-mentioned system.) Now if we call this force simply "the force acting upon the electron,"[13] and maintain the equation[14]

$$\text{mass} \times \text{acceleration} = \text{force}$$

and if we also decide that the accelerations are to be measured in the stationary system K, we derive from the above equations

$$\text{Longitudinal mass} = \frac{m}{\left(\sqrt{1 - v^2/c^2}\right)^3}.$$

$$\text{Transverse mass} = \frac{m}{\sqrt{1 - v^2/c^2}}.$$

With a different definition of force and acceleration we should naturally obtain other values for the masses. This shows us that in comparing different theories of the motion of the electron we must proceed very cautiously.

We remark that these results as to the mass are also valid for ponderable material points,[15] because a ponderable material point can be made into an electron (in our sense of the word) by the addition of an electric charge, *no matter how small.*

We will now determine the kinetic energy of the electron. If an electron moves from rest at the origin of coordinates of the system K along the axis of X under the action of an electrostatic force X, it is clear that the energy withdrawn from the electrostatic field has the value $\int \epsilon X dx$. As the electron is to be slowly accelerated, and consequently may not give off any energy in the form of radiation, the energy withdrawn from the electrostatic field must be put down as equal to the energy of motion W of the electron. Bearing in mind that during the whole process of motion

---

[13]The definition of force here given is not advantageous, as was first shown by M. Planck. It is more to the point to define force in such a way that the laws of momentum and energy assume the simplest form.

[14]EDITOR'S NOTE: Einstein wrote "Massenzahl × Beschleunigungszahl = Kraftzahl", i.e., (numerical value of) mass × (numerical value of) acceleration = (numerical value of) force.

[15]EDITOR'S NOTE: After the publication of this paper Einstein never explicitly stated what he thought of relativistic (velocity-dependent) mass. Hardly in a 1948 letter to Lincoln Barnett [1] he commented on relativistic mass:

> It is not proper to speak of the mass $M = m/(1-v^2/c^2)^{1/2}$ of a moving body, because no clear definition can be given for M. It is preferable to restrict oneself to the "rest mass" m. Besides, one may well use the expression for momentum and energy when referring to the inertial behavior of rapidly moving bodies.

As the quote in [1] is not an exact translation (and appears to suggest that Einstein was explicitly against the concept of relativistic mass M), the translation given here is by Ruschin [2]. A scan of Einstein's letter in German is included in [3].

For the controversy over the concept of relativistic mass in recent years see EDITOR'S NOTE at the end of this article.

[1] See: C. G. Adler, Does mass really depend on velocity, dad? *American Journal of Physics* **55**, 739 (1987)

[2] S. Ruschin, Putting to Rest Mass Misconceptions, *Physics Today*, May 1990, page 15

[3] L. B. Okun, The Concept of Mass, *Physics Today* **42**, 31 (1989)

which we are considering, the first of the equations (A) applies, we therefore obtain

$$W = \int \epsilon X dx = m \int_0^v \beta^3 v \, dv$$

$$= mc^2 \left\{ \frac{1}{\sqrt{1 - v^2/c^2}} - 1 \right\}.$$

Thus, when $v = c$, W becomes infinitely large. Velocities greater than that of light have – as in our previous results – no possibility of existence.

This expression for the kinetic energy must also, by virtue of the argument stated above, apply to ponderable masses as well.

We will now enumerate the properties of the motion of the electron which result from the system of equations (A), and are accessible to experiment.

1. From the second equation of the system (A) it follows that an electric force Y and a magnetic force N have an equally strong deflective action on an electron moving with the velocity v, when $Y = Nv/c$. Thus we see that it is possible by our theory to determine the velocity of the electron from the ratio of the magnetic power of deflection A_m to the electric power of deflection A_e , for any velocity, by applying the law

$$\frac{A_m}{A_e} = \frac{v}{c}.$$

This relationship may be tested experimentally, since the velocity of the electron can be directly measured, e.g. by means of rapidly oscillating electric and magnetic fields.

2. From the deduction for the kinetic energy of the electron it follows that between the potential difference, P, traversed and the acquired velocity v of the electron there must be the relationship

$$P = \int X dx = \frac{m}{\epsilon} c^2 \left\{ \frac{1}{\sqrt{1 - v^2/c^2}} - 1 \right\}.$$

3. We calculate the radius of curvature of the path of the electron when a magnetic force N is present (as the only deflective force), acting perpendicularly to the velocity of the electron. From the second of the equations (A) we obtain

$$-\frac{d^2 y}{dt^2} = \frac{v^2}{R} = \frac{\epsilon}{m} \frac{v}{c} N \sqrt{1 - \frac{v^2}{c^2}}$$

or

$$R = \frac{mc^2}{\epsilon} \frac{v/c}{\sqrt{1 - v^2/c^2}} \frac{1}{N}.$$

These three relationships are a complete expression for the laws according to which, by the theory here advanced, the electron must move.

In conclusion I wish to say that in working at the problem here dealt with I have had the loyal assistance of my friend and colleague M. Besso, and that I am indebted to him for several valuable suggestions.

Bern, June 1905
(Received 30 June 1905)

on the following assertion by Einstein on p. 99:

We remark that these results as to the mass are also valid for ponderable
material points ...

Here Einstein explicitly regarded the mass of ponderable material points as varying with velocity. Since in recent decades there have been attempts to reject the
concept of relativistic (velocity-dependent) mass [1], it should be emphasized that
Einstein's position was and is still correct (despite his doubts revealed in the 1948
letter quoted on p. 99), because it reflects the actual situation in spacetime physics –
both mass and relativistic mass appear to be equally supported by the experimental
evidence. Since mass is defined as the measure of the resistance a particle offers to
its acceleration (which is the accepted definition based on the experimental evidence)
and since it is also an experimental fact that a particle's resistance to its acceleration
increases indefinitely (in a given reference frame) as the particle's velocity approaches
the speed of light (in the same reference frame), it follows that the particle's mass
increases when its velocity increases. Therefore the concept of relativistic mass (like
the concept of mass) reflects an experimental fact.

Some authors point out that $\gamma = 1/\sqrt{1 - \beta^2}$ should not be "attached" to the
mass, because it comes from the 4-velocity. That it comes from the 4-velocity is, of
course, correct – γ ensures that the velocity of a particle cannot exceed that of light;
in other words, γ ensures that no 4-velocity vector, which is timelike, can become
lightlike or spacelike. But that is *kinematics*; it says nothing about *dynamics*, that
is, it says nothing about *why* a particle cannot exceed the velocity of light; in other
words, what is the mechanism that prevents it from doing so? That mechanism is
suggested by Newtonian mechanics, where (to repeat) mass is defined as the measure
of the *resistance* a particle offers to its acceleration; when Einstein postulated that
the velocity of light c is the greatest velocity a particle (with non-zero rest mass) can
achieve, it was almost self-evident to assume that a particle would offer an *increasing
resistance* when accelerated to velocities approaching that of light, that is, a particle's
mass will increase and will approach infinity when the particle's velocity approaches
c. And that was repeatedly experimentally confirmed.

However, increased resistance (i.e., increased relativistic mass) is rather only naming the mechanism that prevents a particle from reaching the velocity of light. The
origin and nature of the resistance a particle offers when accelerated (an open question
in classical physics) and of the *increased* resistance a particle offers when accelerated
to velocities approaching that of light (an open question in spacetime physics) constitute one of the deepest open questions in spacetime physics.

REFERENCES

[1] Unfortunately, the physics community is divided over the concept of relativistic mass. On the one hand, some (mostly particle physicists) firmly reject it (e.g., in
papers entitled "The Virus of Relativistic Mass in the Year of Physics"); indeed during
the last three decades physicists have witnessed (or rather endured) "what has prob-

ably been the most vigorous campaign ever waged against the concept of relativistic mass" [2] . On the other hand, what appears to be the majority of physicists continue to regard it as an integral part of spacetime physics including in books published in recent years. For a more detailed account of the controversy over relativistic mass see the Appendix by the Editor "On Relativistic Mass" in [3].

[2] M. Jammer, *Concepts of Mass in Contemporary Physics and Philosophy* (Princeton University Press, Princeton 2000) p. 51

[3] A. Einstein, *Relativity.* Edited by V. Petkov (Minkowski Institute Press, Montreal 2018)

Does the Inertia of a Body Depend Upon its Energy Content?

A. Einstein

Translation of Ist die Trägheit eines Körpers von seinem Energiegehalt abhängig? *Annalen der Physik*, 17, 1905. Original English publication in: H. A. Lorentz, A. Einstein, H. Minkowski and H. Weyl, *The Principle of Relativity: A Collection of Original Memoirs on the Special and General Theory of Relativity*. With Notes by A. Sommerfeld. Translated by W. Perrett and G. B. Jeffery (Methuen and Company, Ltd., 1923; reprinted by Dover Publications Inc., 1952).

The results of the previous investigation lead to a very interesting conclusion, which is here to be deduced.

I based that investigation on the Maxwell-Hertz equations for empty space, together with the Maxwellian expression for the electromagnetic energy of space, and in addition the principle that:

The laws by which the states of physical systems alter are independent of the alternative, to which of two systems of coordinates, in uniform motion of parallel translation relatively to each other, these alterations of state are referred (principle of relativity).

With these principles[16] as my basis I deduced *inter alia* the following result (§ 8):

Let a system of plane waves of light, referred to the system of coordinates x, y, z), possess the energy l; let the direction of the ray (the wave-normal) make an angle ϕ with the axis of x of the system. If we introduce a new system of coordinates (ξ, η, ζ) moving in uniform parallel translation with respect to the system (x, y, z), and having its origin of coordinates in motion along the axis of x with the velocity v then this quantity of light – measured in the system (ξ, η, ζ) – possesses the energy

$$l^* = l \, \frac{1 - \dfrac{v}{c} \cos \phi}{\sqrt{1 - \dfrac{v^2}{c^2}}}$$

where c denotes the velocity of light. We shall make use of this result in what follows.

Let there be a stationary body in the system (x, y, z), and let its energy – referred

---

[16]The principle of the constancy of the velocity of light is of course contained in Maxwell's equations.

to the system (x, y, z) – be E_0 . Let the energy of the body relative to the system (ξ, η, ζ), moving as above with the velocity v, be H_0.

Let this body send out, in a direction making an angle ϕ with the axis of x plane waves of light, of energy $\frac{1}{2}L$ measured relatively to (x, y, z), and simultaneously an equal quantity of light in the opposite direction. Meanwhile the body remains at rest with respect to the system (x, y, z). The principle of energy must apply to this process, and in fact (by the principle of relativity) with respect to both systems of coordinates. If we call the energy of the body after the emission of light E_1 or H_1 respectively, measured relatively to the system (x, y, z) or (ξ, η, ζ) respectively, then by employing the relation given above we obtain

$$E_0 = E_1 + \frac{L}{2} + \frac{L}{2},$$

$$H_0 = H_1 + \frac{L}{2} \frac{1 - \frac{v}{c}\cos\phi}{\sqrt{1 - \frac{v^2}{c^2}}} - \frac{L}{2} \frac{1 + \frac{v}{c}\cos\phi}{\sqrt{1 - \frac{v^2}{c^2}}}$$

$$= H_1 + \frac{L}{\sqrt{1 - \frac{v^2}{c^2}}}.$$

By subtraction we obtain from these equations

$$(H_0 - E_0) - (H_1 - E_1) = L \left\{ \frac{1}{\sqrt{1 - \frac{v^2}{c^2}}} - 1 \right\}.$$

The two differences of the form $H - E$ occurring in this expression have simple physical significations. H and E are energy values of the same body referred to two systems of coordinates which are in motion relatively to each other, the body being at rest in one of the two systems (system (x, y, z)). Thus it is clear that the difference $H - E$ can differ from the kinetic energy K of the body, with respect to the other system (ξ, η, ζ), only by an additive constant C, which depends on the choice of the arbitrary additive constants of the energies H and E. Thus we may place

$$H_0 - E_0 = K_0 + C,$$
$$H_1 - E_1 = K_1 + C,$$

since C does not change during the emission of light. So we have

$$K_0 - K_1 = L \left\{ \frac{1}{\sqrt{1 - \frac{v^2}{c^2}}} - 1 \right\}.$$

The kinetic energy of the body with respect to (ξ, η, ζ) diminishes as a result of the emission of light, and the amount of diminution is independent of the properties of the body. Moreover, the difference $K_0 - K_1$, like the kinetic energy of the electron (§ 10), depends on the velocity.

Neglecting magnitudes of fourth and higher orders we may place

$$K_0 - K_1 = \frac{L}{2}\frac{v^2}{c^2}.$$

From this equation it directly follows that:

If a body gives off the energy L in the form of radiation, its mass diminishes by L/c^2. The fact that the energy withdrawn from the body becomes energy of radiation evidently makes no difference, so that we are led to the more general conclusion that:

The mass of a body is a measure of its energy-content; if the energy changes by L, the mass changes in the same sense by $L/9 \times 10^{20}$, the energy being measured in ergs, and the mass in grammes.

It is not impossible that with bodies whose energy-content is variable to a high degree (e.g. with radium salts) the theory may be successfully put to the test.

If the theory corresponds to the facts, radiation conveys inertia between the emitting and absorbing bodies.

SPACE AND TIME

H. MINKOWSKI

Lecture[1] given at the 80th Meeting of the Natural Scientists in Cologne on September 21, 1908.

New translation (by V. Petkov) in: Hermann Minkowski, *Spacetime: Minkowski's Papers on Spacetime Physics*. Translated by Gregorie Dupuis-Mc Donald, Fritz Lewertoff and Vesselin Petkov. Edited by V. Petkov (Minkowski Institute Press, Montreal 2020), pp. 57-76.

Gentlemen! The views of space and time which I want to present to you arose from the domain of experimental physics, and therein lies their strength. Their tendency is radical. From now onwards space by itself and time by itself shall completely fade into mere shadows and only a specific union of the two will still stand independently on its own.

I.

I want to show first how to move from the currently adopted mechanics through purely mathematical reasoning to modified ideas about space and time. The equations of Newtonian mechanics show a twofold invariance. First, their form is preserved when subjecting the specified spatial coordinate system to *any change of position*; second, when it changes its state of motion, namely when any *uniform translation* is impressed upon it; also, the zero point of time plays no role. When one feels ready for the axioms of mechanics, one is accustomed to regard the axioms of geometry as settled and probably for this reason those two invariances are rarely mentioned in the same breath. Each of them represents a certain group of transformations for the differential equations of mechanics. The existence of the first group can be seen as reflecting a fundamental characteristic of space. One always tends to treat the second group with disdain in order to unburden one's mind that one can never determine from physical phenomena whether space, which is assumed to be at rest, may not after all be in uniform translation. Thus these two groups lead completely separate

---

[1]H. Minkowski, Raum und Zeit, *Physikalische Zeitschrift* **10** (1909) S. 104-111; *Jahresbericht der Deutschen Mathematiker-Vereinigung* **18** (1909) S. 75-88; reprinted in *Gesammelte Abhandlungen von Hermann Minkowski*, ed. by D. Hilbert, 2 vols. (Teubner, Leipzig 1911), vol. 2, pp. 431-444, and in H.A. Lorentz, A. Einstein, H. Minkowski, Das Relativitätsprinzip (Teubner, Leipzig 1913) S. 56-68. This lecture also appeared as a separate publication (booklet): H. Minkowski, *Raum und Zeit* (Teubner, Leipzig 1909).

lives side by side. Their entirely heterogeneous character may have discouraged any intention to compose them. But it is the composed complete group as a whole that gives us to think.

We will attempt to visualize the situation graphically. Let x, y, z be orthogonal coordinates for space and let t denote time. The objects of our perception are always connected to places and times. No one has noticed a place other than at a time and a time other than at a place. However I still respect the dogma that space and time each have an independent meaning. I will call a point in space at a given time, i.e. a system of values x, y, z, t a *worldpoint*. The manifold of all possible systems of values x, y, z, t will be called the *world*. With a hardy piece of chalk I can draw four world axes on the blackboard. Even *one* drawn axis consists of nothing but vibrating molecules and also makes the journey with the Earth in the Universe, which already requires sufficient abstraction; the somewhat greater abstraction associated with the number 4 does not hurt the mathematician. To never let a yawning emptiness, let us imagine that everywhere and at any time something perceivable exists. In order not to say matter or electricity I will use the word substance for that thing. We focus our attention on the substantial point existing at the worldpoint x, y, z, t and imagine that we can recognize this substantial point at any other time. A time element dt may correspond to the changes dx, dy, dz of the spatial coordinates of this substantial point. We then get an image, so to say, of the eternal course of life of the substantial point, a curve in the world, a *worldline*, whose points can be clearly related to the parameter t from $-\infty$ to $+\infty$. The whole world presents itself as resolved into such worldlines, and I want to say in advance, that in my understanding the laws of physics can find their most complete expression as interrelations between these worldlines.

Through the concepts of space and time the x, y, z-manifold $t = 0$ and its two sides $t > 0$ and $t < 0$ fall apart. If for simplicity we hold the chosen origin of space and time fixed, then the first mentioned group of mechanics means that we can subject the x, y, z-axes at $t = 0$ to an arbitrary rotation about the origin corresponding to the homogeneous linear transformations of the expression

$$x^2 + y^2 + z^2.$$

The second group, however, indicates that, also without altering the expressions of the laws of mechanics, we may replace

$$x, y, z, t \quad \text{by} \quad x - \alpha t, \ y - \beta t, \ z - \gamma t, \ t,$$

where α, β, γ are any constants. The time axis can then be given a completely arbitrary direction in the upper half of the world $t > 0$. What has now the requirement of orthogonality in space to do with this complete freedom of choice of the direction of the time axis upwards?

To establish the connection we take a positive parameter c and look at the structure

$$c^2 t^2 - x^2 - y^2 - z^2 = 1.$$

It consists of two sheets separated by $t = 0$ by analogy with a two-sheeted hyperboloid. We consider the sheet in the region $t > 0$ and we will now take those

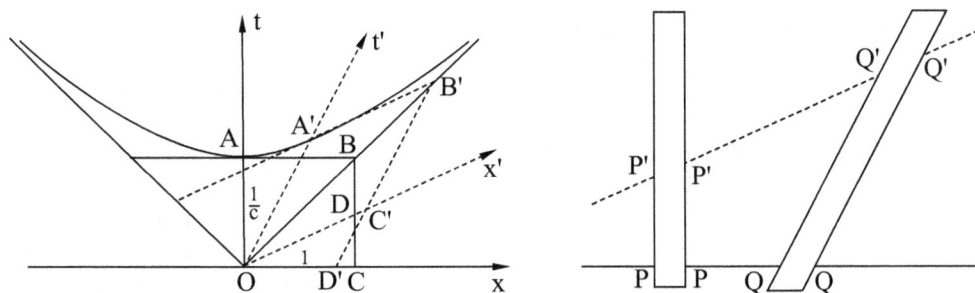

Fig. 1

homogeneous linear transformations of x, y, z, t in four new variables x', y', z', t' so that the expression of this sheet in the new variables has the same form. Obviously, the rotations of space about the origin belong to these transformations. A full understanding of the rest of those transformations can be obtained by considering such among them for which y and z remain unchanged. We draw (Fig. 1) the intersection of that sheet with the plane of the x- and the t-axis, i.e. the upper branch of the hyperbola $c^2 t^2 - x^2 = 1$ with its asymptotes. Further we draw from the origin O an arbitrary radius vector OA' of this branch of the hyperbola; then we add the tangent to the hyperbola at A' to intersects the right asymptote at B'; from $OA'B'$ we complete the parallelogram $OA'B'C'$; finally, as we will need it later, we extend $B'C'$ so that it intersects the x-axis at D'. If we now regard OC' and OA' as axes for new coordinates x', t', with the scale units $OC' = 1$, $OA' = 1/c$, then that branch of the hyperbola again obtains the expression $ct'^2 - x'^2 = 1$, $t' > 0$, and the transition from x, y, z, t to x', y', z', t' is one of the transformations in question. These transformations plus the arbitrary displacements of the origin of space and time constitute a group of transformations which still depends on the parameter c and which I will call G_c.

   If we now increase c to infinity, so $1/c$ converges to zero, it is clear from the figure that the branch of the hyperbola leans more and more towards the x-axis, that the angle between the asymptotes becomes greater, and in the limit that special transformation converts to one where the t'-axis may be in any upward direction and x' approaches x ever more closely. By taking this into account it becomes clear that the group G_c in the limit $c = \infty$, that is the group G_∞, is exactly the complete group which is associated with the Newtonian mechanics. In this situation, and since G_c is mathematically more understandable than G_∞, there could have probably been a mathematician with a free imagination who could have come up with the idea that at the end natural phenomena do not actually possess an invariance with the group G_∞, but rather with a group G_c with a certain finite c, which is *extremely great* only in the ordinary units of measurement. Such an insight would have been an extraordinary triumph for pure mathematics. Now mathematics expressed only staircase wit here, but it has the satisfaction that, due to its happy antecedents with their senses sharpened by their free and penetrating imagination, it can grasp the profound consequences of such remodelling of our view of nature.

I want to make it quite clear what the value of c will be with which we will be finally dealing. c is the *velocity of the propagation of light in empty space*. To speak neither of space nor of emptiness, we can identify this magnitude with the ratio of the electromagnetic to the electrostatic unit of the quantity of electricity.

The existence of the invariance of the laws of nature with respect to the group G_c would now be stated as follows:

From the entirety of natural phenomena, through successively enhanced approximations, it is possible to deduce more precisely a reference system x, y, z, t, space and time, by means of which these phenomena can be then represented according to certain laws. But this reference system is by no means unambiguously determined by the phenomena. *One can still change the reference system according to the transformations of the above group G_c arbitrarily without changing the expression of the laws of nature in the process.*

For example, according to the figure depicted above one can call t' time, but then must necessarily, in connection with this, define space by the manifold of three parameters x', y, z in which the laws of physics would then have exactly the same expressions by means of x', y, z, t' as by means of x, y, z, t. Hereafter we would then have in the world no more *the* space, but an infinite number of spaces analogously as there is an infinite number of planes in three-dimensional space. Three-dimensional geometry becomes a chapter in four-dimensional physics. You see why I said at the beginning that space and time will recede completely to become mere shadows and only a world in itself will exist.

II.

Now the question is, what circumstances force us to the changed view of space and time, does it actually never contradict the phenomena, and finally, does it provide advantages for the description of the phenomena?

Before we discuss these questions, an important remark is necessary. Having individualized space and time in some way, a straight worldline parallel to the t-axis corresponds to a stationary substantial point, a straight line inclined to the t-axis corresponds to a uniformly moving substantial point, a somewhat curved worldline corresponds to a non-uniformly moving substantial point. If at any worldpoint x, y, z, t there is a worldline passing through it and we find it parallel to any radius vector OA' of the previously mentioned hyperboloidal sheet, we may introduce OA' as a new time axis, and with the thus given new concepts of space and time, the substance at the worldpoint in question appears to be at rest. We now want to introduce this fundamental axiom:

With appropriate setting of space and time the substance existing at any worldpoint can always be regarded as being at rest.

This axiom means that at every worldpoint[2] the expression

$$c^2 \mathrm{d}t^2 - \mathrm{d}x^2 - \mathrm{d}y^2 - \mathrm{d}z^2$$

is always positive, which is equivalent to saying that any velocity v is always smaller than c. Then c would be an upper limit for all substantial velocities and that is

---

[2] *Editor's note:* Minkowski means at every worldpoint along the worldline of the substance.

precisely the deeper meaning of the quantity c. In this understanding the axiom is at first glance slightly displeasing. It should be noted, however, that a modified mechanics, in which the square root of that second order differential expression enters, is now gaining ground, so that cases with superluminal velocity will play only such a role as that of figures with imaginary coordinates in geometry.

The *impulse* and true motivation for *accepting the group* G_c came from noticing that the differential equation for the propagation of light waves in the empty space possesses that group G_c[3]. On the other hand, the concept of a rigid body has meaning only in a mechanics with the group G_∞. If one has optics with G_c, and if, on the other hand, there were rigid bodies, it is easy to see that *one* t-direction would be distinguished by the two hyperboloidal sheets corresponding to G_c and G_∞, and would have the further consequence that one would be able, by using appropriate rigid optical instruments in the laboratory, to detect a change of phenomena at various orientations with respect to the direction of the Earth's motion. All efforts directed towards this goal, especially a famous interference experiment of Michelson had, however, a negative result. To obtain an explanation, H. A. Lorentz made a hypothesis, whose success lies precisely in the invariance of optics with respect to the group G_c. According to Lorentz every body moving at a velocity v must experience a reduction in the direction of its motion namely in the ratio

$$1 : \sqrt{1 - \frac{v^2}{c^2}}.$$

This hypothesis sounds extremely fantastical. Because the contraction is not to be thought of as a consequence of resistances in the ether, but merely as a gift from above, as an accompanying circumstance of the fact of motion.

I now want to show on our figure that the Lorentzian hypothesis is completely equivalent to the new concept of space and time, which makes it much easier to understand. If for simplicity we ignore y and z and think of a world of one spatial dimension, then two strips, one upright parallel to the t-axis and the other inclined to the t-axis (see Fig. 1), are images for the progression in time of a body at rest and a body moving uniformly, where each preserves a constant spatial dimension. OA' is parallel to the second strip, so we can introduce t' as time and x' as a space coordinate and then it appears that the second body is at rest, whereas the first – in uniform motion. We now assume that the first body has length l when considered at rest, that is, the cross section PP of the first strip and the x-axis is equal to $l \cdot OC$, where OC is the measuring unit on the x-axis, and, on the other hand, that the second body has the same length l when *regarded at rest*; then the latter means that the cross-section of the second strip *measured parallel to the x'-axis* is $Q'Q' = l \cdot OC'$. We have now in these two bodies images of two *equal* Lorentz electrons, one stationary and one uniformly moving. But if we go back to the original coordinates x, t, we should take as the dimension of the second electron the cross section QQ of its associated strip *parallel to the x-axis*. Now as $Q'Q' = l \cdot OC'$, it is obvious that $QQ = l \cdot OD'$. If

---

[3]An important application of this fact can already be found in W. Voigt, Göttinger Nachrichten, 1887, S. 41.

dx/dt for the second strip is $= v$, an easy calculation gives

$$OD' = OC \cdot \sqrt{1 - \frac{v^2}{c^2}},$$

therefore also $PP : QQ = 1 : \sqrt{1 - \frac{v^2}{c^2}}$. This is the meaning of the Lorentzian hypothesis of the contraction of electrons in motion. Regarding, on the other hand, the second electron as being at rest, that is, adopting the reference system x', t', the length of the first electron will be the cross section $P'P'$ of its strip parallel to OC', and we would find the first electron shortened with respect to the second in exactly the same proportion; from the figure we also see that

$$P'P' : Q'Q' = OD : OC' = OD' : OC = QQ : PP.$$

Lorentz called t', which is a combination of x and t, *local time* of the uniformly moving electron, and associated a physical construction with this concept for a better understanding of the contraction hypothesis. However, it is to the credit of A. Einstein[4] who first realized clearly that the time of one of the electrons is as good as that of the other, i.e. that t and t' should be treated equally. With this, time was deposed from its status as a concept unambiguously determined by the phenomena. Neither Einstein nor Lorentz disputed the concept of space, maybe because in the above-mentioned special transformation, where the plane of x', t' coincides with the plane x, t, an interpretation is possible as if the x-axis of space preserved its position. To go beyond the concept of space in such a way is an instance of what can only be imputed to the audacity of mathematical culture. After this further step, which is indispensable for the true understanding of the group G_c, I think the word *relativity postulate* used for the requirement of invariance under the group G_c is very feeble. Since the meaning of the postulate is that through the phenomena only the four-dimensional world in space and time is given, but the projection in space and in time can still be made with certain freedom, I want to give this affirmation rather the name *the postulate of the absolute world* (or shortly the world postulate).

III.

Through the world postulate an identical treatment of the four identifying quantities x, y, z, t becomes possible. I want to explain now how, as a result of this, we gain more understanding of the forms under which the laws of physics present themselves. Especially the concept of *acceleration* acquires a sharply prominent character.

I will use a geometric way of expression, which presents itself immediately when one implicitly ignores z in the triple x, y, z. An arbitrary worldpoint O can be taken as the origin of space-time. The *cone*

$$c^2 t^2 - x^2 - y^2 - z^2 = 0$$

with O as the apex (Fig. 2) consists of two parts, one with values $t < 0$, the other one with values $t > 0$.

---

[4]A. Einstein, Annalen der Physik 17 (1905), S. 891; Jahrbuch der Radioaktivität und Elektronik 4 (1907), S. 411.

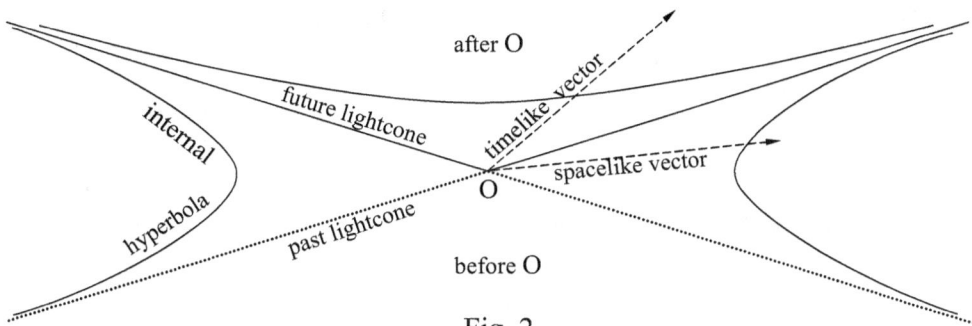

Fig. 2

The first, the *past lightcone of O*, consists, we can say, of all worldpoints which "send light to O", the second, the *future lightcone of O*, consists of all worldpoints which "receive light from O"[5]. The area bounded solely by the past lightcone may be called *before O*, whereas the area bounded solely by the future lightcone – *after O*. Situated after O is the already considered hyperboloidal sheet

$$F = c^2t^2 - x^2 - y^2 - z^2 = 1, \ t > 0$$

The area *between the cones* is filled with the one-sheeted hyperboloidal structures

$$-F = x^2 + y^2 + z^2 - c^2t^2 = k^2$$

for all constant positive values of k^2. Essential for us are the hyperbolas with O as the center, located on the latter structures. The individual branches of these hyperbolas may be briefly called *internal hyperbolas with center O*. Such a hyperbola would be thought of as the worldline of a substantive point, which represents its motion that increases asymptotically to the velocity of light c for $t = -\infty$ and $t = +\infty$.

If we now call, by analogy with vectors in space, a directed line in the manifold x, y, z, t a vector, we have to distinguish between the *timelike* vectors with directions from O to the sheet $+F = 1, t > 0$, and the *spacelike* vectors with directions from O to $-F = 1$. The time axis can be parallel to any vector of the first kind. Every worldpoint between the future lightcone and the past lightcone of O can be regarded, by a choice of the reference system, as *simultaneous* with O as well as *earlier* than O or *later* than O. Each worldpoint within the past lightcone of O is necessarily always earlier than O, each worldpoint within the future lightcone is necessarily always later than O. The transition to the limit $c = \infty$ would correspond to a complete folding of the wedge-shaped section between the cones into the flat manifold $t = 0$. In the figures this section is intentionally made with different widths.

---

[5] *Editor's and translator's note:* I decided to translate the words *Vorkegel* and *Nachkegel* as *past lightcone* and *future lightcone*, respectively, for two reasons. First, this translation reflects the essence of Minkowski's idea – (i) all worldpoints on the past lightcone "send *light* to O", which means that they all can influence O and therefore lie in the past of O; (ii) all worldpoints on the future lightcone "receive *light* from O", which means that they all can be influenced by O and therefore lie in the *future* of O. Second, the terms *past lightcone* and *future lightcone* are now widely accepted in spacetime physics.

We decompose any vector, such as that from O to x, y, z, t into four *components* x, y, z, t. If the directions of two vectors are, respectively, that of a radius vector OR from O to one of the surfaces $\mp F = 1$, and that of a tangent RS at the point R on the same surface, the vectors are called *normal* to each other. Accordingly,

$$c^2 t t_1 - x x_1 - y y_1 - z z_1 = 0$$

is the condition for the vectors with components x, y, z, t and x_1, y_1, z_1, t_1 to be normal to each other.

The *measuring units* for the *magnitudes* of vectors in different directions may be fixed by assigning to a spacelike vector from O to $-F = 1$ always the magnitude 1, and to a timelike vector from O to $+F = 1, t > 0$ always the magnitude $1/c$.

Let us now imagine a worldpoint $P(x, y, z, t)$ through which the worldline of a substantial point is passing, then the magnitude of the timelike vector dx, dy, dz, dt along the line will be

$$d\tau = \frac{1}{c}\sqrt{c^2 dt^2 - dx^2 - dy^2 - dz^2}.$$

The integral $\int d\tau = \tau$ of this magnitude, taken along the worldline from any fixed starting point P_0 to the variable end point P, we call the *proper time* of the substantial point at P. On the worldline we consider x, y, z, t, i.e. the components of the vector OP, as functions of the proper time τ; denote their first derivatives with respect to τ by $\dot{x}, \dot{y}, \dot{z}, \dot{t}$; their second derivatives with respect to τ by $\ddot{x}, \ddot{y}, \ddot{z}, \ddot{t}$, and call the corresponding vectors, the derivative of the vector OP with respect to τ the *velocity vector at P* and the derivative of the velocity vector with respect to τ the *acceleration vector at P*. As

$$c^2 \dot{t}^2 - \dot{x}^2 - \dot{y}^2 - \dot{z}^2 = c^2$$

it follows that

$$c^2 \dot{t}\ddot{t} - \dot{x}\ddot{x} - \dot{y}\ddot{y} - \dot{z}\ddot{z} = 0,$$

i.e. the velocity vector is the timelike vector of magnitude 1 in the direction of the worldline at P, and the acceleration vector at P is normal to the velocity vector at P, so it is certainly a spacelike vector.

Now there is, as is easily seen, a specific branch of the hyperbola, which has three infinitely adjacent points in common with the worldline at P, and whose asymptotes are generators of a past lightcone and a future lightcone (see Fig. 3). This branch of the hyperbola will be called the *curvature hyperbola* at P. If M is the center of this hyperbola, we have here an internal hyperbola with center M. Let ρ be the magnitude of the vector MP, *so we recognize the acceleration vector at P as the vector in the direction MP of magnitude c^2/ρ.*

If $\ddot{x}, \ddot{y}, \ddot{z}, \ddot{t}$ are all zero, the curvature hyperbola reduces to the straight line touching the worldline at P, and we should set $\rho = \infty$.

IV.

To demonstrate that the adoption of the group G_c for the laws of physics never leads to a contradiction, it is inevitable to undertake a revision of all physics based

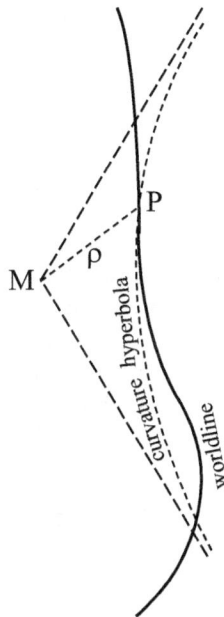

Fig. 3

on the assumption of this group. This revision has been done successfully to some extent for questions of thermodynamics and heat radiation[6], for the electromagnetic processes, and finally, with the retention of the concept of mass, for mechanics.[7]

For the latter domain, the question that should be raised above all is: When a force with the spatial components X, Y, Z acts at a worldpoint $P(x, y, z, t)$, where the velocity vector is $\dot{x}, \dot{y}, \dot{z}, \dot{t}$, as what force this force should be interpreted for any change of the reference system? Now there exist some proven approaches to the ponderomotive force in the electromagnetic field in cases where the group G_c is undoubtedly permissible. These approaches lead to the simple rule: *When the reference system is changed, the given force transforms into a force in the new space coordinates in such a way that the corresponding vector with the components*

$$\dot{t}X, \ \dot{t}Y, \ \dot{t}Z, \ \dot{t}T$$

remains unchanged, and where

$$T = \frac{1}{c^2}\left(\frac{\dot{x}}{\dot{t}}X + \frac{\dot{y}}{\dot{t}}Y + \frac{\dot{z}}{\dot{t}}Z\right)$$

is the work done by the force at the worldpoint divided by c^2. This vector is always

---

[6]M. Planck, "Zur Dynamik bewegter Systeme," Sitzungsberichte der k. preußischen Akademie der Wissenschaften zu Berlin, 1907, S. 542 (auch Annalen der Physik, Bd. 26, 1908, S. 1).

[7]H. Minkowski, "Die Grundgleichungen für die elektromagnetischen Vorgänge in bewegten Körpern", Nachrichten der k. Gesellschaft der Wissenschaft zu Göttingen, mathematisch-physikalische Klasse, 1908, S. 53 und Mathematische Annalen, Bd. 68, 1910, S. 527

normal to the velocity vector at P. Such a force vector, representing a force at P, will be called a *motive force vector* at P.

Now let the worldline passing through P represent a substantial point with constant *mechanical mass m*. The multiplied by m velocity vector at P will be called the *momentum vector at P*, and the multiplied by m acceleration vector at P will be called the *force vector of the motion at P*. According to these definitions, the law of motion for a point mass with a given force vector is:[8]

The force vector of the motion is equal to the motive force vector.

This assertion summarizes four equations for the components for the four axes, wherein the fourth can be regarded as a consequence of the first three because both vectors are from the start normal to the velocity vector. According to the above meaning of T, the fourth equation is undoubtedly the law of conservation of energy. The *kinetic energy* of the point mass is defined as the *component of the momentum vector along the t-axis multiplied by* c^2. The expression for this is

$$mc^2 \frac{dt}{d\tau} = \frac{mc^2}{\sqrt{1 - \frac{v^2}{c^2}}},$$

which is, the expression $\frac{1}{2}mv^2$ of Newtonian mechanics after the subtraction of the additive constant term mc^2 and neglecting magnitudes of the order $1/c^2$. The *dependence of the energy on the reference system* is manifested very clearly here. But since the t-axis can be placed in the direction of each timelike vector, then, on the other hand, the law of conservation of energy, formed for every possible reference system, already contains the whole system of the equations of motion. In the discussed limiting case $c = \infty$, this fact will retain its importance for the axiomatic structure of Newtonian mechanics and in this sense has been already noticed by J. R. Schütz[9]

From the beginning we can determine the ratio of the units of length and time in such a way that the natural limit of velocity becomes $c = 1$. If we introduce $\sqrt{-1}t = s$ instead of t, then the quadratic differential expression

$$d\tau^2 = -dx^2 - dy^2 - dz^2 - ds^2$$

becomes completely symmetric in x, y, z, s and this symmetry is carried over to any law that does not contradict the world postulate. Thus the essence of this postulate can be expressed mathematically very concisely in the mystical formula:

$$3 \cdot 10^5 \text{ km} = \sqrt{-1} \text{ seconds.}$$

V.

The advantages resulting from the world postulate may most strikingly be proved by indicating the effects from *an arbitrarily moving point charge* according to the

---

[8]H. Minkowski, loc. cit., p. 107. Cf. also M. Planck, Verhandlungen der Physikalischen Gesellschaft, Bd. 4, 1906, S. 136.

[9]J. R. Schütz, "Das Prinzip der absoluten Erhaltung der Energie", Nachrichten der k. Gesellschaft der Wissenschaften zu Göttingen, mathematisch-physikalische Klasse, 1897, S. 110.

Maxwell-Lorentz theory. Let us imagine the worldline of such a pointlike electron with charge e, and take on it the proper time τ from any initial point. To determine the field induced by the electron at any worldpoint P_1 we construct the past lightcone corresponding to P_1 (Fig. 4). It intersects the infinite worldline of the electron obviously at a single point P because the tangents to every point on the worldline are all timelike vectors. At P we draw the tangent to the worldline and through P_1 construct the normal P_1Q to this tangent. Let the magnitude of P_1Q be r. According to the definition of a past lightcone the magnitude of PQ should be r/c. *Now the vector of magnitude e/r in the direction PQ represents through its components along the x-, y-, z-axes, the vector potential multiplied by c, and through the component along the t-axis, the scalar potential of the field produced by e at the worldpoint P_1.* This is the essence of the elementary laws formulated by A. Liénard and E. Wiechert.[10]

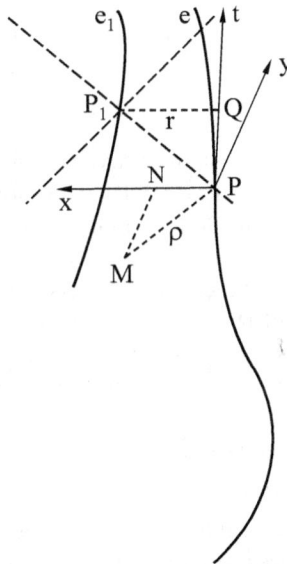

Fig. 4

Then it emerges in the description itself of the field caused by the electron that the division of the field into electric and magnetic forces is a relative one with respect to the specified time axis; most clearly the two forces considered together can be described in some, though not complete, analogy with the wrench in mechanics. I now want to describe *the ponderomotive action of an arbitrarily moving point charge exerted on another arbitrarily moving point charge.* Let us imagine that the worldline of a second pointlike electron of charge e_1 goes through the worldpoint P_1. We define P, Q, r as before, then construct (Fig. 4) the center M of the curvature hyperbola at P, and finally the normal MN from M to an imagined straight line from P parallel to QP_1. We now fix a reference system with its origin at P in the following way: the

[10]A. Liénard, "Champ électrique et magnétique produit par une charge concentré en un point et animée d'un mouvement quelconque", L'Éclairage électrique, T. 16, 1898, pp. 5, 53, 106; E. Wiechert, "Elektrodynamische Elementargesetze", Archives Néerlandaiaes des Sciences exactes et naturelles (2), T. 5, 1900, S. 549.

118

t-axis in the direction of PQ, the x-axis in the direction of QP_1, the y-axis in the direction of MN, and lastly the direction of the z-axis is determined as being normal to the t-, x-, y-axes. Let the acceleration vector at P be $\ddot{x}, \ddot{y}, \ddot{z}, \ddot{t}$, the velocity vector at P_1 be $\dot{x}_1, \dot{y}_1, \dot{z}_1, \dot{t}_1$. *Now the motive force vector exerted by the first arbitrarily moving electron e on the second arbitrarily moving electron e_1 at P_1 will be*

$$-ee_1(\dot{t}_1 - \frac{\dot{x}_1}{c})\mathfrak{K}$$

where for the components $\mathfrak{K}_x, \mathfrak{K}_y, \mathfrak{K}_z, \mathfrak{K}_t$ of the vector $\mathfrak{K}$ three relations exist:

$$c\mathfrak{K}_t - \mathfrak{K}_x = \frac{1}{r^2}, \quad \mathfrak{K}_y = \frac{\ddot{y}}{c^2 r}, \quad \mathfrak{K}_z = 0$$

and fourthly this vector $\mathfrak{K}$ is normal to the velocity vector at P_1, and this circumstance alone makes it dependent on the latter velocity vector.

If we compare this assertion with the previous formulations[11] of the same elementary law of the ponderomotive action of moving point charges on one another, we are compelled to admit that the relations considered here reveal their inner being in full simplicity only in four dimensions, whereas on a three dimensional space, forced upon us from the beginning, they cast only a very tangled projection.

In mechanics reformed in accordance with the world postulate, the disturbing disharmony between Newtonian mechanics and the modern electrodynamics disappears by itself. In addition, I want to touch on the status of the *Newtonian law of attraction* with respect to this postulate. I will consider two point masses m, m_1, represented by their worldlines, and that m exerts a motive force vector on m_1 exactly as in the case of electrons, except that instead of $-ee_1 + mm_1$ should be used. We can now specifically consider the case when the acceleration vector of m is constantly zero, then we may choose t in such a way that m is regarded as at rest, and assume that only m_1 move under the motive force vector which originates from m. If we now modify this specified vector by adding the factor $\dot{t}^{-1} = \sqrt{1 - \frac{v^2}{c^2}}$, which up to magnitudes of the order $1/c^2$ is equal to 1, it can be seen[12] that for the positions x_1, y_1, z_1 of m_1 and their progression in time, we arrive exactly at Kepler's laws, except that instead of the times t_1 the proper times τ_1 of m_1 should be used. On the basis of this simple remark we can then see that the proposed law of attraction associated with the new mechanics is no less well suited to explain the astronomical observations than the Newtonian law of attraction associated with the Newtonian mechanics.

The fundamental equations for the electromagnetic processes in ponderable bodies are entirely in accordance with the world postulate. Actually, as I will show elsewhere, there is no need to abandon the derivation of these equations which is based on ideas of the electron theory as taught by Lorentz.

The validity without exception of the world postulate is, I would think, the true core of an electromagnetic world view which, as Lorentz found it and Einstein further

[11]K. Schwazschild, Nachrichten der k. Gesellschaft der Wissenschaften zu Göttinger, mathematisch-physikalische Klasse, 1903, S. 132; H. A. Lorentz, Enzyklopädie der mathematischen Wissenschaften, V, Art. 14, S. 199.
[12]H. Minkowski, loc. cit., p. 110.

unveiled it, lies downright and completely exposed before us as clear as daylight. With the development of the mathematical consequences of this postulate, sufficient findings of its experimental validity will be arrived at so that even those to whom it seems unsympathetic or painful to abandon the prevailing views become reconciled through the thought of a pre-stabilized harmony between mathematics and physics.

NOTES
by
A. Sommerfeld

Original English publication in: H. A. Lorentz, A. Einstein, H. Minkowski and H. Weyl, *The Principle of Relativity: A Collection of Original Memoirs on the Special and General Theory of Relativity.* With Notes by A. Sommerfeld. Translated by W. Perrett and G. B. Jeffery (Methuen and Company, Ltd., 1923; reprinted by Dover Puplications Inc., 1952).

The following notes are given in an appendix so as to interfere in no way with Minkowski's text. They are by no means essential, having no other purpose than that of removing certain small formal mathematical difficulties which might hinder the comprehension of Minkowski's great thoughts. The bibliographical references are confined to the literature dealing expressly with the subject of his address. From the physical point of view there is nothing in what Minkowski says that must now be withdrawn, with the exception of the final remark on Newton's law of attraction. What will be the epistemological attitude towards Minkowski's conception of the time-space problem is another question, but, as it seems to me, a question which does not essentially touch his physics.

(1) Page 111.[1] "On the other hand, the concept of rigid bodies has meaning only in mechanics satisfying the group G_∞." This sentence was confirmed in the widest sense in a discussion on a paper by his disciple M. Born, a year after Minkowski's death. Born (Ann. d. Physik, 30, 1909, p. 1) had defined a relatively rigid body as one in which every element of volume, even in accelerated motions, undergoes the Lorentzian contraction appropriate to its velocity. Ehrenfest (Phys. Zeitschr., 10, 1909, p. 918) showed that such a body cannot be set in rotation; Herglotz (Ann. d. Phys., 31, 1910, p. 393) and F. Nöther (Ann. d. Phys., 31, 1910, p. 919) that it has only three degrees of freedom of movement. The attempt was also made to define a relatively rigid body with six or nine degrees of freedom. But Planck (Phys. Zeitschr., 11, 1910, p. 294) expressed the view that the theory of relativity can operate only with more or less elastic bodies, and Laue (Phys. Zeitschr., 12, 1911, p. 48), employing Minkowski's methods, and his Fig. 2 in the text above, proved that in the theory of relativity every solid body must have an infinite number of degrees of

---

[1] EDITOR'S NOTE: The quotes here are from the 1923 translation of Minkowski's paper by W. Perrett and G. B. Jeffery.

freedom. Finally Herglotz (Ann. d. Physik, 36, 1911, p. 453) developed a relativistic theory of elasticity, according to which elastic tensions always occur if the motion of the body is not relatively rigid in Born's sense. Thus the relatively rigid body plays the same part in this theory of elasticity as the ordinary rigid body plays in the ordinary theory of elasticity.

(2) Page 112. "If dx/dt for the second band is equal to v, an easy calculation gives $OD' = OC\sqrt{1 - v^2/c^2}$." In Fig. 1, let $\alpha = \angle A'OA, \beta = \angle B'OA' = \angle C'OB'$, in which the equality of the last two angles follows from the symmetrical position of the asymptotes with respect to the new axes of coordinates (conjugate diameters of the hyperbola).[2] Since $\alpha + \beta = \frac{1}{4}\pi$,

$$\sin 2\beta = \cos 2\alpha.$$

In the triangle $OD'C'$ the law of sines gives

$$\frac{OD'}{OC'} = \frac{\sin 2\beta}{\cos \alpha} = \frac{\cos 2\alpha}{\cos \alpha}$$

or, as $OC' = OA'$,

$$OD' = OA' \frac{\cos 2\alpha}{\cos \alpha} = OA' \cos \alpha(1 - \tan^2 \alpha). \tag{1}$$

If x, t are the coordinates of the point A' in the x, t system, and therefore $x.OA$ and $ct.OC = ct.OA$ respectively are the corresponding distances from the axes of coordinates, we have

$$x.OA = \sin \alpha.OA', \quad ct.OA = \cos \alpha.OA', \quad \frac{x}{ct} = \tan \alpha = \frac{v}{c}. \tag{2}$$

Inserting these values of x and ct in the equation of the hyperbola, we find

$$OA'^2(\cos^2 \alpha - \sin^2 \alpha) = OA^2, \quad OA' = \frac{OA}{\cos \alpha\sqrt{1 - \tan^2 \alpha}}. \tag{3}$$

therefore, on account of (1) and (2),

$$OD' = OA\sqrt{1 - \tan^2 \alpha} = OA\sqrt{1 - v^2/c^2}.$$

This, because $OA = OC$, is the formula to be proved.

Further, in the right-angled triangle OCD,

$$OD = \frac{OC}{\cos \alpha} = \frac{OA}{\cos \alpha}.$$

Equation (3) may therefore be also written in this way,

$$OA' = \frac{OD}{\sqrt{1 - \tan^2 \alpha}} \quad \text{or} \quad \frac{OD}{OA'} = \sqrt{1 - \frac{v^2}{c^2}}.$$

---

[2] Sommerfeld seems to take ct as a coordinate in the graph in place of t as used by Minkowski.— Note by the translators (W. Perrett and G. B. Jeffery).

This, together with (4), gives the proportion,

$$OD : OA' = OD' : OA,$$

which, as $OA' = OC'$ and $OA = OC$, is identical with

$$OD : OC' = OD' : OC$$

employed on page 112.

(3) Page 113. "Any world-point between the front and back cones of O can be arranged, by means of the system of reference, so as to be simultaneous with O, but also just as well so as to be earlier than O, or later than O." M. Laue (Phys. Zeitschr., 12, 1911, p. 48) traces to this observation the proof of Einstein's theorem: In the theory of relativity no process of causality can be propagated with a velocity greater than that of light ("Signal velocity $\leqq c$"). Assume that an event O causes another event P, and that the world-point P lies in the region between the cones of O. In this case the effect would have been conveyed from O to P with a velocity greater than that of light, relatively to the system of references x, t in question, in which, of course, the effect P is assumed to be later than the cause O, $t_P > 0$. But now, in accordance with the words quoted above, the system of reference may be changed, so that P comes to be earlier than O, that is to say, a system x', t' may be chosen in infinitely many ways so that t'_P becomes < 0. This is irreconcilable with the idea of causality. P must therefore lie either "after" O or on the back cone of O, i.e. the velocity of propagation of a signal to be sent from O, which is to cause a second event at the world-point P, must of necessity be $\leqq c$. Of course it is possible, even in the theory of relativity, to define processes propagated with velocity greater than light. This can be done geometrically, for example, in a very simple way. But such processes can never serve as signals, i.e. it is impossible to introduce them arbitrarily and by them, for example, to set a relay in motion at a distant place. There may be e.g. optical media, in which the "velocity of light" is greater than c. But in that case what is understood by the velocity of light is the propagation of phases in an infinite periodic wave-train. These can never be used for signalling. On the other hand a wave-front is propagated, in all circumstances and with any constitution of the optical medium, with the velocity c; cf. e.g. A. Sommerfeld, "Festschrift Heinrich Weber," Leipzig, Teubner, 1912, p. 338, or Ann. d. Physik, 44, 1914, p. 177.

(4) Page 114. As Minkowski once remarked to me, the element of proper time $d\tau$ is not a complete differential. Thus if we connect two world-points O and P by two different world-lines 1 and 2, then

$$\int_1 d\tau \neq \int_2 d\tau.$$

If 1 runs parallel to the t-axis, so that the first transition in the chosen system of reference signifies rest, it is evident that

$$\int_1 d\tau = t, \qquad \int_2 d\tau < t.$$

On this depends the retardation of the moving clock compared with the clock at rest. The assertion is based, as Einstein has pointed out, on the unprovable assumption that the clock in motion actually indicates its own proper time, i.e. that it always gives the time corresponding to the state of velocity, regarded as constant, at any instant. The moving clock must naturally have been moved with acceleration (with changes of speed or direction) in order to be compared with the stationary clock at the world-point P. The retardation of the moving clock does not therefore actually indicate "motion," but "accelerated motion." Hence this does not contradict the principle of relativity.

(5) Page 114. The term "hyperbola of curvature" is formed exactly on the model of the elementary concept of the circle of curvature. The analogy becomes analytical identity if instead of the real coordinate of time t the imaginary $u = ict$ is employed, that is, c times the coordinate employed by Minkowski, page 116.

By page 113 an internal hyperbola in the x, t -plane has the equation, with $k = \rho$,

$$x^2 - c^2 t^2 = \rho^2,$$

therefore in the x, u plane

$$x^2 + u^2 = \rho^2.$$

Hence it may be written in parametric form, when ϕ denotes a purely imaginary angle,

$$x = \rho \cos \phi, \qquad u = \rho \sin \phi.$$

So, as I suggested in the Ann. d. Phys., 33, p. 649, § 8, hyperbolic motion may also be denoted as "cyclic motion," whereby its chief properties (convection of the field, occurrence of a kind of centrifugal force) are characterized with particular clearness. For the hyperbolic motion we have

$$d\tau = \frac{1}{c}\sqrt{-du^2 - dx^2} = \frac{\rho}{c}[d\phi]$$

and thus

$$\dot{x} = \frac{dx}{d\tau} = -ic \sin \phi, \qquad \dot{u} = \frac{du}{d\tau} = ic \cos \phi$$

$$\ddot{x} = \frac{d\dot{x}}{d\tau} = \frac{c^2}{\rho} \cos \phi, \qquad \ddot{u} = \frac{d\dot{u}}{d\tau} = \frac{c^2}{\rho} \sin \phi.$$

The magnitude of the acceleration vector in hyperbolic motion is therefore c^2/ρ. Since any given world-line is touched by the hyperbola of curvature at three points, it has the same acceleration vector as the hyperbolic motion, and its magnitude is c^2/ρ, as indicated on page 114.

The centre M of the cyclic motion $x^2 + u^2 = \rho^2$ is evidently the point $x = 0, u = 0$, and from this centre all points of the hyperbola have the constant "distance," i.e. a constant magnitude of the radius vector. Therefore ρ denotes the interval marked MP in Fig. 3.

(6) Page 115. A force X, Y, Z, to be made into a "force vector," must be multiplied by $\dot{t} = dt/d\tau$. This may be explained as follows.

According to Minkowski, page 115, the momentum vector is defined by $m\dot{x}$, $m\dot{y}$, $m\dot{z}$, $m\dot{t}$, where m denotes the "constant mechanical mass," or, as Minkowski says more plainly elsewhere, the "rest mass." If we retain Newton's law of motion (time rate of change of momentum equals to force), we have to set

$$\frac{d}{dt}(m\dot{x}) = X, \quad \frac{d}{dt}(m\dot{y}) = Y, \quad \frac{d}{dt}(m\dot{z}) = Z.$$

Multiplication by i makes the left-hand sides into vector components in Minkowski's sense. Therefore iX, iY, iZ are also the first three components of the "force vector." The fourth component T follows without ambiguity from the requirement that the force vector is to be normal to the motion vector. Minkowski's equations for the mechanics of the mass point are therefore, with constant rest mass,

$$m\ddot{x} = iX, \quad m\ddot{y} = iY, \quad m\ddot{z} = iZ, \quad m\ddot{t} = iT.$$

The assumption of constancy of rest mass can only be maintained, however, when the energy-content of the body is not changed in its motion, or in the words of Planck, when the motion ensues "adiabatically and isochorically."

(7) Pages 117. What is characteristic of the constructions here given, is their complete independence of any special system of reference. They give, as Minkowski postulates on page 108, "reciprocal relations between world-lines" (or world-points) as " the most perfect expression of physical laws." On page 117, for example, the electrodynamic potential (four-potential) is not referred to the axes of coordinates x, y, z, t until it is to be conventionally divided into a scalar and a vector portion, which have no independent invariant meaning from the relativistic standpoint.

By way of commentary to Minkowski I have deduced, from Maxwell's equations, by Minkowski's methods, an invariant analytical form for the four-potential and the ponderomotive action between two electrons, and so given another view of these constructions of Minkowski. Instead of going into details here, I may refer to my article in Ann. d. Phys., 33, 1910, p. 649, § 7, or to M. Laue, "Das Relativitatsprinzip," Braunschweig, Vieweg, 1913, § 19. Compare also Minkowski's address on the principle of relativity, edited by myself, in Ann. d. Phys., 47, 1915, p. 927, where the four-potential is placed at the head of electrodynamics, and this theory thus reduced to its simplest form.

(8) Page 117. The invariant representation of the electromagnetic field by a "vector of the second kind" (or, as I proposed to call it, a "six-vector," a term which seems to be winning acceptance) is a particularly important part of Minkowski's view of electrodynamics. Whereas Minkowski's ideas on the vector of the first kind, or four-vector, were in part anticipated by Poincaré (Rend. Circ. Mat. Palermo, 21, 1906), the introduction of the six-vector is new. Like the six-vector, the wrench of mechanics (standing for a single force and a couple) depends on six independent parameters. And as in the electromagnetic field "the separation into electric and magnetic force is a relative one," so with the wrench, as is well known, the division into single force and couple can be made in very many ways.

(9) Page 118. Minkowski's relativistic form of Newton's law for the special case of zero acceleration mentioned in the text is included in the more general form proposed

by Poincaré (loc. cit.). On the other hand, in taking acceleration into consideration, it goes further than the latter. Minkowski's or Poincaré's formulation of the law of gravitation shows that it is possible in many ways to reconcile Newton's law with the theory of relativity. That law is viewed as a point law, and gravitation therefore in a certain sense as action at a distance. The general theory of relativity, which Einstein has been developing from 1907 on, gets a deeper grip of the problem of gravitation. Gravitation is not only regarded as a field action and described by space-time differential equations – which seems from the present standpoint irrefutable – but it is also united organically with the principle of relativity extended to any transformations, whereas Minkowski and Poincaré had adapted it to the postulate of relativity in a more external manner. In the general theory of relativity the space-time structure is determined, from or together with, gravitation. Thus the principle of relativity, by an extension of Minkowski's ideas, is so formulated that it postulates the covariance of physical quantities with reference to all point transformations, so that the coefficients of the invariant linear element enter into the laws of physics.

(10) Page 118. The "fundamental equations for electromagnetic processes in ponderable bodies" are developed by Minkowski in Göttinger Nachrichten, 1907. It was not granted him to complete the "deduction of this equation on the basis of the theory of electrons." His essays in this direction have been worked out by M. Born, and together with the "Fundamental Equations" make up the first volume of the series of monographs edited by Otto Blumenthal (Leipzig, 1910).

Part II

GRAVITY IS BUT GEOMETRY

ON THE INFLUENCE OF GRAVITATION ON THE PROPAGATION OF LIGHT

A. EINSTEIN

Translation of Über den Einfluss der Schwerkraft auf die Ausbreitung des Lichtes, *Annalen der Physik*, 35, 1911. Original English publication in: H. A. Lorentz, A. Einstein, H. Minkowski and H. Weyl, *The Principle of Relativity: A Collection of Original Memoirs on the Special and General Theory of Relativity*. With Notes by A. Sommerfeld. Translated by W. Perrett and G. B. Jeffery (Methuen and Company, Ltd., 1923; reprinted by Dover Publications Inc., 1952).

In a memoir published four years ago[1] I tried to answer the question whether the propagation of light is influenced by gravitation. I return to this theme, because my previous presentation of the subject does not satisfy me, and for a stronger reason, because I now see that one of the most important consequences of my former treatment is capable of being tested experimentally. For it follows from the theory here to be brought forward, that rays of light, passing close to the sun, are deflected by its gravitational field, so that the angular distance between the sun and a fixed star appearing near to it is apparently increased by nearly a second of arc.

In the course of these reflections further results are yielded which relate to gravitation. But as the exposition of the entire group of considerations would be rather difficult to follow, only a few quite elementary reflections will be given in the following pages, from which the reader will readily be able to inform himself as to the suppositions of the theory and its line of thought. The relations here deduced, even if the theoretical foundation is sound, are valid only to a first approximation.

§ 1. A Hypothesis as to the Physical Nature of the Gravitational Field

In a homogeneous gravitational field (acceleration of gravity γ) let there be a stationary system of coordinates K, orientated so that the lines of force of the gravitational field run in the negative direction of the axis of z. In a space free of gravitational fields let there be a second system of coordinates K', moving with uniform acceleration (7) in the positive direction of its axis of z. To avoid unnecessary complications, let us for the present disregard the theory of relativity, and regard both systems from the customary point of view of kinematics, and the movements

---

[1] A. Einstein, Jahrbuch für Radioakt. und Elektronik, 4, 1907.

occurring in them from that of ordinary mechanics.

Relatively to K, as well as relatively to K', material points which are not subjected to the action of other material points, move in keeping with the equations

$$\frac{d^2x}{dt^2} = 0, \qquad \frac{d^2y}{dt^2} = 0, \qquad \frac{d^2z}{dt^2} = -\gamma.$$

For the accelerated system K' this follows directly from Galileo's principle, but for the system K, at rest in a homogeneous gravitational field, from the experience that all bodies in such a field are equally and uniformly accelerated. This experience, of the equal falling of all bodies in the gravitational field, is one of the most universal which the observation of nature has yielded; but in spite of that the law has not found any place in the foundations of our edifice of the physical universe.

But we arrive at a very satisfactory interpretation of this law of experience, if we assume that the systems K and K' are physically exactly equivalent, that is, if we assume that we may just as well regard the system K as being in a space free from gravitational fields, if we then regard K as uniformly accelerated. This assumption of exact physical equivalence makes it impossible for us to speak of the absolute acceleration[2] of the system of reference, just as the usual theory of relativity forbids

---

[2]EDITOR'S NOTE: Apparently, in 1911 Einstein had not completely overcome his initial hostile attitude toward Minkowski's demonstration that what Einstein called (special) *relativity* is in fact a theory of an *absolute* four-dimensional world (spacetime) in which particles are a forever given web of worldlines (or rather worldtubes): "Since the mathematicians have invaded the relativity theory, I do not understand it myself any more" (Albert Einstein quoted by A. Sommerfeld, "To Albert Einstein's Seventieth Birthday." In: *Albert Einstein: Philosopher-Scientist*. P. A. Schilpp, ed., 3rd ed. (Open Court, Illinois 1969) pp. 99-105, p. 102). Otherwise, he would have taken full advantage of Minkowski's 1908 lecture "Space and Time" (this volume) where he pointed out that acceleration in spacetime is absolute since it is represented by an *absolute* geometrical property of the worldline of an accelerating particle – its curvature, i.e., its deformation. In Minkowski's words (this volume, p. 110): "a somewhat curved worldline corresponds to a non-uniformly moving substantial point" and "Especially the concept of *acceleration* acquires a sharply prominent character" (this volume, p. 112) namely because it reflects an *absolute* geometrical property (the curvature, i.e., the deformation) of the worldline of an accelerating particle.

Had Einstein started to study Minkowski's works earlier, he might have realized that *there is nothing genuinely relative in his theory of relativity*, might have adopted earlier the four-dimensional mathematical formalism of spacetime (the *absolute* four-dimensional world introduced by Minkowski) and might have probably found a different path to the idea that gravitational phenomena are merely manifestations of the non-Euclidean geometry of spacetime.

Einstein might have been impressed by the link, discovered by Minkowski, between the experimental fact that a particle *resists* its acceleration and the geometrical fact that the worldtube of an accelerating particle is *curved* or *deformed*. Then Einstein would have certainly made use of the fact he used in the actual development of general relativity – that all particles fall toward the Earth with the same acceleration regardless of their masses – and his famous thought experiments, particularly the one analyzing physical phenomena in a lift on the Earth's surface and in an accelerating lift, would have led him to the conclusion that *a falling body does not resist its fall*.

Then the path to the idea that gravitational phenomena are manifestations of the curvature of spacetime would have been open to Einstein – the experimental fact that a falling particle accelerates (which means that its worldtube is curved), but offers no resistance to its acceleration (which means that its worldtube is not deformed) can be explained only if the worldtube of a falling particle is both curved and not deformed, which is impossible in the flat Minkowski spacetime where a curved worldtube is always deformed. Such a worldtube can exist only in a non-Euclidean spacetime whose geodesic worldtubes are naturally curved due to the spacetime curvature, but are not deformed.

us to talk of the absolute velocity of a system;[3] and it makes the equal falling of all bodies in a gravitational field seem a matter of course.

As long as we restrict ourselves to purely mechanical processes in the realm where Newton's mechanics holds sway, we are certain of the equivalence of the systems K and K'. But this view of ours will not have any deeper significance unless the systems K and K' are equivalent with respect to all physical processes, that is, unless the laws of nature with respect to K are in entire agreement with those with respect to K'. By assuming this to be so, we arrive at a principle which, if it is really true, has great heuristic importance. For by theoretical consideration of processes which take place relatively to a system of reference with uniform acceleration, we obtain information as to the career of processes in a homogeneous gravitational field. We shall now show, first of all, from the standpoint of the ordinary theory of relativity, what degree of probability is inherent in our hypothesis.

§ 2. On the Gravitation of Energy

One result yielded by the theory of relativity is that the inertial mass of a body increases with the energy it contains; if the increase of energy amounts to E, the increase in inertial mass is equal to E/c^2, when c denotes the velocity of light. Now is there an increase of gravitating mass corresponding to this increase of inertial mass? If not, then a body would fall in the same gravitational field with varying acceleration according to the energy it contained. That highly satisfactory result of the theory of relativity by which the law of the conservation of mass is merged in the law of conservation of energy could not be maintained, because it would compel us to abandon the law of the conservation of mass in its old form for inertial mass, and maintain it for gravitating mass.

But this must be regarded as very improbable. On the other hand, the usual theory of relativity does not provide us with any argument from which to infer that the weight of a body depends on the energy contained in it. But we shall show that our hypothesis of the equivalence of the systems K and K' gives us gravitation of energy as a necessary consequence.

Let the two material systems S_1 and S_2, provided with instruments of measurement, be situated on the z-axis of K at the distance h from each other,[4] so that the gravitation potential in S_2 is greater than that in S_1 by γh. Let a definite quantity of energy E be emitted from S_2 towards S_1. Let the quantities of energy in S_1 and S_2 be measured by contrivances which – brought to one place in the system z and there compared – shall be perfectly alike. As to the process of this conveyance of energy by radiation we can make no *a priori* assertion, because we do not know the influence of the gravitational field on the radiation and the measuring instruments in S_1 and S_2.

But by our postulate of the equivalence of K and K' we are able, in place of the system K in a homogeneous gravitational field, to set the gravitation-free sys-

---

[3]Of course we cannot replace any arbitrary gravitational field by a state of motion of the system without a gravitational field, any more than, by a transformation of relativity, we can transform all points of a medium in any kind of motion to rest.

[4]The dimensions of S_1 and S_2 are regarded as infinitely small in comparison with h.

tem K', which moves with uniform acceleration in the direction of positive z and with the z-axis of which the material systems S_1 and S_2 are rigidly connected.

We judge of the process of the transference of energy by radiation from S_2 to S_1 from a system K_0, which is to be free from acceleration. At the moment when the radiation energy E_2 is emitted from S_2 toward S_1, let the velocity of K' relatively to K_0 be zero. The radiation will arrive at S_1 when the time h/c has elapsed (to a first approximation). But at this moment the velocity of S_1 relatively to K_0 is $\gamma\, h/c = v$. Therefore by the ordinary theory of relativity the radiation arriving at S_1 does not possess the energy E_2, but a greater energy E_1, which is related to E_2 to a first approximation by the equation

$$E_1 = E_2\left(1 + \frac{v}{c}\right) = E_2\left(1 + \gamma\frac{h}{c^2}\right) \qquad (1)$$

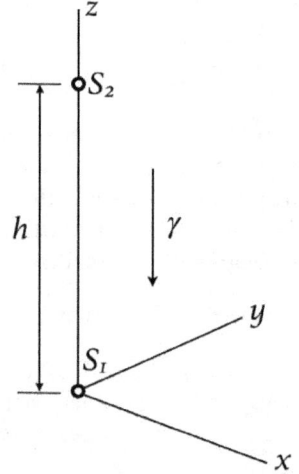

Fig. 1

By our assumption exactly the same relation holds if the same process takes place in the system K, which is not accelerated, but is provided with a gravitational field. In this case we may replace γh by the potential Φ of the gravitation vector in S_2, if the arbitrary constant of Φ in S_1 is equated to zero. We then have the equation

$$E_1 = E_2 + \frac{E_2}{c^2}\Phi. \qquad (1a)$$

This equation expresses the law of energy for the process under observation. The energy E_1 arriving at S_1 is greater than the energy E_2, measured by the same means, which was emitted in S_2, the excess being the potential energy of the mass E_2/c^2 in the gravitational field. It thus proves that for the fulfilment of the principle of energy we have to ascribe to the energy E, before its emission in S_2, a potential energy due to gravity, which corresponds to the gravitational mass E/c^2. Our assumption of the equivalence of K and K' thus removes the difficulty mentioned at the beginning of this paragraph which is left unsolved by the ordinary theory of relativity.

The meaning of this result is shown particularly clearly if we consider the following cycle of operations:

1. The energy E, as measured in S_2, is emitted in the form of radiation in S_2 towards S_1, where, by the result just obtained, the energy $E(1 + \gamma h/c)$, as measured in S_1, is absorbed.

2. A body W of mass M is lowered from S_2 to S_1, work $M\gamma h$ being done in the process.

3. The energy E is transferred from S_1 to the body W while W is in S_1. Let the gravitational mass M be thereby changed so that it acquires the value M'.

4. Let W be again raised to S_2, work $M'\gamma h$ being done in the process.

5. Let E be transferred from W back to S_2.

The effect of this cycle is simply that S_1 has undergone the increase of energy $E\gamma h/c^2$, and that the quantity of energy $M'\gamma h - M\gamma h$ has been conveyed to the system in the form of mechanical work. By the principle of energy, we must therefore have

$$E\gamma\frac{h}{c^2} = M'\gamma h - M\gamma h,$$

or

$$M' - M = E/c^2. \tag{1b}$$

The increase in gravitational mass is thus equal to E/c^2, and therefore equal to the increase in inertial mass as given by the theory of relativity.

The result emerges still more directly from the equivalence of the systems K and K' according to which the gravitational mass in respect of K is exactly equal to the inertial mass in respect of K'; energy must therefore possess a gravitational mass which is equal to its inertial mass. If a mass M_0 be suspended on a spring balance in the system K', the balance will indicate the apparent weight $M_0\gamma$ on account of the inertia of M_0. If the quantity of energy E be transferred to M_0, the spring balance, by the law of the inertia of energy, will indicate $(M_0 + E/c^2)\gamma$. By reason of our fundamental assumption exactly the same thing must occur when the experiment is repeated in the system K, that is, in the gravitational field.

§ 3. Time and the Velocity of Light in the Gravitational Field

If the radiation emitted in the uniformly accelerated system K' in S_2 toward S_1 had the frequency ν_2 relatively to the clock in S_2, then, relatively to S_1, at its arrival in S_1 it no longer has the frequency ν_2 relatively to an identical clock in S_1, but a greater frequency ν_1 lt such that to a first approximation

$$\nu_1 = \nu_2 \left(1 + \gamma\frac{h}{c^2}\right). \tag{2}$$

For if we again introduce the unaccelerated system of reference K_0, relatively to which, at the time of the emission of light, K' has no velocity, then S_1, at the time of arrival of the radiation at S_1, has, relatively to K_0, the velocity $\gamma h/c$, from which, by Doppler's principle, the relation as given results immediately.

In agreement with our assumption of the equivalence of the systems K' and K, this equation also holds for the stationary system of coordinates K, provided with a uniform gravitational field, if in it the transference by radiation takes place as described. It follows, then, that a ray of light emitted in S_2 with a definite gravitational potential, and possessing at its emission the frequency ν_2 – compared with a clock in S_2 – will, at its arrival in S_1, possess a different frequency ν_1 – measured by an identical clock in S_1. For γh we substitute the gravitational potential Φ of S_2 – that of S_1 being taken as zero – and assume that the relation which we have deduced for the homogeneous gravitational field also holds for other forms of field. Then

$$\nu_1 = \nu_2 \left(1 + \frac{\Phi}{c^2}\right).\tag{2a}$$

This result (which by our deduction is valid to a first approximation) permits, in the first place, of the following application. Let ν_0 be the vibration-number of an elementary light-generator, measured by a delicate clock at the same place. Let us imagine them both at a place on the surface of the Sun (where our S_2 is located). Of the light there emitted, a portion reaches the Earth (S_1), where we measure the frequency of the arriving light with a clock U in all respects resembling the one just mentioned. Then by (2a),

$$\nu = \nu_0 \left(1 + \frac{\Phi}{c^2}\right).$$

where Φ is the (negative) difference of gravitational potential between the surface of the Sun and the Earth. Thus according to our view the spectral lines of sunlight, as compared with the corresponding spectral lines of terrestrial sources of light, must be somewhat displaced toward the red, in fact by the relative amount

$$\frac{\nu_0 - \nu}{\nu_0} = -\frac{\Phi}{c^2} = 2.10^{-6}.$$

If the conditions under which the solar bands arise were exactly known, this shifting would be susceptible of measurement. But as other influences (pressure, temperature) affect the position of the centres of the spectral lines, it is difficult to discover whether the inferred influence of the gravitational potential really exists.[5]

On a superficial consideration equation (2), or (2a), respectively, seems to assert an absurdity. If there is constant transmission of light from S_2 to S_1; how can any other number of periods per second arrive in S_1 than is emitted in S_2? But the answer is simple. We cannot regard ν_2 or respectively ν_1 simply as frequencies (as the number of periods per second) since we have not yet determined the time in system K. What ν_2 denotes is the number of periods with reference to the time-unit of the clock U in S_2, while ν_1 denotes the number of periods per second with reference to the identical clock in S_1. Nothing compels us to assume that the clocks U in different gravitation potentials must be regarded as going at the same rate. On the contrary, we must certainly define the time in K in such a way that the number of wave crests and troughs between S_2 and S_1 is independent of the absolute value of time; for the process under observation is by nature a stationary one. If we did not satisfy this condition, we should arrive at a definition of time by the application of which time would merge explicitly into the laws of nature, and this would certainly be unnatural and unpractical. Therefore the two clocks in S_1 and S_2 do not both give the "time" correctly. If we measure time in S_1 with the clock U, then we must measure time in S_2 with a clock which goes $(1 + \Phi/c^2)$ times more slowly than the

---

[5]L. F. Jewell (Journ. de Phys., 6, 1897, p. 84) and particularly Ch. Fabry and H. Boisson (Comptes rendus, 148, 1909, pp. 688-690) have actually found such displacements of fine spectral lines toward the red end of the spectrum, of the order of magnitude here calculated, but have ascribed them to an effect of pressure in the absorbing layer.

clock U when compared with U at one and the same place. For when measured by such a clock the frequency of the ray of light which is considered above is at its emission in S_2

$$\nu_2 \left(1 + \frac{\Phi}{c^2}\right).$$

and is therefore, by (2a), equal to the frequency ν_1 of the same ray of light on its arrival in S_1.

This has a consequence which is of fundamental importance for our theory. For if we measure the velocity of light at different places in the accelerated, gravitation-free system K', employing clocks U of identical constitution, we obtain the same magnitude at all these places. The same holds good, by our fundamental assumption, for the system K as well. But from what has just been said we must use clocks of unlike constitution, for measuring time at places with differing gravitation potential. For measuring time at a place which, relatively to the origin of the coordinates, has the gravitation potential Φ, we must employ a clock which – when removed to the origin of coordinates – goes $(1 + \Phi/c^2)$ times more slowly than the clock used for measuring time at the origin of coordinates. If we call the velocity of light at the origin of coordinates c_0, then the velocity of light c at a place with the gravitation potential Φ will be given by the relation

$$c = c_0 \left(1 + \frac{\Phi}{c^2}\right). \tag{3}$$

The principle of the constancy of the velocity of light holds good according to this theory in a different form from that which usually underlies the ordinary theory of relativity.

§ 4. Bending of Light-Rays in the Gravitational Field

From the proposition which has just been proved, that the velocity of light in the gravitational field is a function of the place, we may easily infer, by means of Huyghens's principle, that light-rays propagated across a gravitational field undergo deflection. For let E be a wave front of a plane light-wave at the time t, and let P_1 and P_2 be two points in that plane at unit distance from each other. P_1 and P_2 lie in

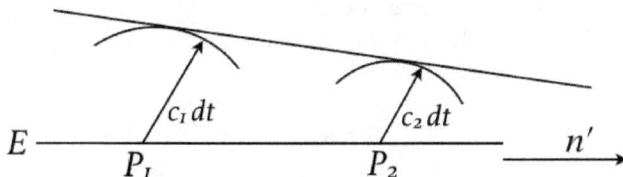

Fig. 2

the plane of the paper, which is chosen so that the differential coefficient of Φ taken in the direction of the normal to the plane, vanishes, and therefore also that of c. We obtain the corresponding wave front at time $t + dt$, or, rather, its line of section with the plane of the paper, by describing circles round the points P_1 and P_2 with radii $c_1 dt$ and $c_2 dt$ respectively, where c_1 and c_2 denote the velocity of light at the points P_1 and P_2 respectively, and by drawing the tangent to these circles. The angle through which the light-ray is deflected in the path cdt is therefore

$$(c_1 - c_2)dt = -\frac{\partial c}{\partial n'}\, dt,$$

if we calculate the angle positively when the ray is bent toward the side of increasing n'. The angle of deflection per unit of path of the light-ray is thus

$$-\frac{1}{c}\frac{\partial c}{\partial n'}, \quad \text{or by (3)} \quad -\frac{1}{c^2}\frac{\partial \Phi}{\partial n'}$$

Finally, we obtain for the deflection which a light-ray experiences toward the side n' on any path(s) the expression

$$\alpha = -\frac{1}{c^2}\int \frac{\partial \Phi}{\partial n'}\, ds. \tag{4}$$

We might have obtained the same result by directly considering the propagation of a ray of light in the uniformly accelerated system K', and transferring the result to the system K, and thence to the case of a gravitational field of any form.

By equation (4) a ray of light passing along by a heavenly body suffers a deflection to the side of the diminishing gravitational potential, that is, on the side directed toward the heavenly body, of the magnitude

$$\alpha = \frac{1}{c^2}\int_{\theta=-\frac{1}{2}\pi}^{\theta=\frac{1}{2}\pi} \frac{kM}{r^2}\cos\theta\, ds = 2\frac{kM}{c^2\Delta},$$

where k denotes the constant of gravitation, M the mass of the heavenly body, Δ the distance of the ray from the centre of the body. A ray of light going past the Sun would accordingly undergo deflection to the amount of $4 \cdot 10^{-6} = 0 \cdot 83$ seconds of arc. The angular distance of the star from the centre of the Sun appears to be increased by this amount. As the fixed stars in the parts of the sky near the Sun are visible during total eclipses of the Sun, this consequence of the theory may be compared with experience. With the planet Jupiter the displacement to be expected reaches to about $\frac{1}{100}$ of the amount given. It would be a most desirable thing if astronomers would take up the question here raised. For apart from any theory there is the question whether it is possible with the equipment at present available to detect an influence of gravitational fields on the propagation of light.

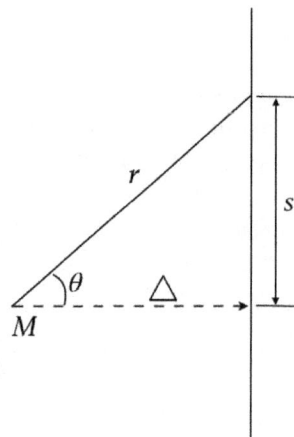

Fig. 3

Prague, June 1911

THE FOUNDATION OF THE GENERAL THEORY OF RELATIVITY

A. EINSTEIN

Translation of Die Grundlage der allgemeinen Relativitätstheorie, *Annalen der Physik*, 49, 1916. Original English publication in: H. A. Lorentz, A. Einstein, H. Minkowski and H. Weyl, *The Principle of Relativity: A Collection of Original Memoirs on the Special and General Theory of Relativity*. With Notes by A. Sommerfeld. Translated by W. Perrett and G. B. Jeffery (Methuen and Company, Ltd., 1923; reprinted by Dover Publications Inc., 1952).

EDITOR'S NOTE: The beginning of the paper (the first page before section "A. Fundamental Considerations on the Postulate of Relativity"), where Einstein acknowledged the role of his mathematics professor Hermann Minkowski in preparing the ground for his general relativity (by developing the four-dimensional formalism of the introduced by him spacetime structure of the world), was omitted in the 1923 English publication in *The Principle of Relativity*. The reason seems to be that that publication had been a direct translation (without checking the original publication of Einstein's paper in *Annalen der Physik*) of the German collection *Das Relativitätsprinzip: Eine Samm-lung von Abhandlungen*[1] and the first page of Einstein papers had been somehow omitted. The missing page was translated by André Michaud and Fritz Lewertoff and is included below.

The theory presented in the following pages is the most far-reaching generalization of the theory currently named the "Theory of Relativity;" the latter I will name "Special Relativity" in the following pages to distinguish it from the generalized form with the assumption that it is known. The generalization of the theory of relativity was greatly facilitated by the form given to special relativity by Minkowski, who was the first mathematician to clearly recognize the formal equivalence of the spatial coordinates and the time coordinate, which made it useful for the construction of the theory. The mathematical tools necessary for general relativity were already provided by the "absolute differential calculus," which is based on the research of Gauss, Riemann and Christoffel on non-Euclidean manifolds and was integrated into a system by Ricci and Levi-Civita that was already applied to problems in theoretical

---

[1]H. A. Lorentz, A. Einstein, H. Minkowski, *Das Relativitätsprinzip: Eine Sammlung von Ab-handlungen*. Mit einem Beitrag von H. Weyl und Anmerkungen von A. Sommerfeld. Vorwort von O. Blumenthal, Fünfte Auflage (Springer Fachmedien Wiesbaden Gmbh, 1923)

136

physics. In Section B of the present treatise I have developed all the mathematical aids that are required and that cannot be assumed to be known by physicists in the simplest and most transparent way possible, so that a study of mathematical literature is not required to understand the present treatise. Finally, I would like to thank my friend, the mathematician Grossmann, who not only saved me from studying the relevant mathematical literature, but also supported me in my search for the field equations of gravitation.

A. FUNDAMENTAL CONSIDERATIONS ON THE POSTULATE OF RELATIVITY

§ 1. Observations on the Special Theory of Relativity

The special theory of relativity is based on the following postulate, which is also satisfied by the mechanics of Galileo and Newton.

If a system of coordinates K is chosen so that, in relation to it, physical laws hold good in their simplest form, the same laws also hold good in relation to any other system of coordinates K' moving in uniform translation relatively to K. This postulate we call the "special principle of relativity." The word "special" is meant to intimate that the principle is restricted to the case when K' has a motion of uniform translation relatively to K, but that the equivalence of K' and K does not extend to the case of non-uniform motion of K' relatively to K.

Thus the special theory of relativity does not depart from classical mechanics through the postulate of relativity, but through the postulate of the constancy of the velocity of light in vacuo , from which, in combination with the special principle of relativity, there follow, in the well-known way, the relativity of simultaneity, the Lorentzian transformation, and the related laws for the behaviour of moving bodies and clocks.

The modification to which the special theory of relativity has subjected the theory of space and time is indeed far-reaching, but one important point hs remained unaffected. For the laws of geometry, even according to the special theory of relativity, are to be interpreted directly as laws relating to the possible relative positions of solid bodies at rest; and, in a more general way, the laws of kinematics are to be interpreted as laws which describe the relations of measuring bodies and clocks. To two selected material points of a stationary rigid body there always corresponds a distance of quite definite length, which is independent of the locality and orientation of the body, and is also independent of the time. To two selected positions of the hands of a clock at rest relatively to the privileged system of reference there always corresponds an interval of time of a definite length, which is independent of place and time. We shall soon see that the general theory of relativity cannot adhere to this simple physical interpretation of space and time.

§ 2. The Need for an Extension of the Postulate of Relativity

In classical mechanics, and no less in the special theory of relativity, there is an inherent epistemological defect which was, perhaps for the first time, clearly pointed out by Ernst Mach. We will elucidate it by the following example: Two fluid bodies of the same size and nature hover freely in space at so great a distance from each other and from all other masses that only those gravitational forces need be taken into account which arise from the interaction of different parts of the same body. Let the distance between the two bodies be invariable, and in neither of the bodies let there be any relative movements of the parts with respect to one another. But let either mass, as judged by an observer at rest relatively to the other mass, rotate with constant angular velocity about the line joining the masses. This is a verifiable relative motion of the two bodies. Now let us imagine that each of the bodies has been surveyed by means of measuring instruments at rest relatively to itself, and let the surface of S_1 prove to be a sphere, and that of S_2 an ellipsoid of revolution. Thereupon we put the question – What is the reason for this difference in the two bodies? No answer can be admitted as epistemoiogicany satisfactory,[2] unless tne reason given is an *observable fact of experience*. The law of causality has not the significance of a statement as to the world of experience, except when *observable facts* ultimately appear as causes and effects.

Newtonian mechanics does not give a satisfactory answer to this question. It pronounces as follows: The laws of mechanics apply to the space R_1 in respect to which the body S_1 is at rest, but not to the space R_2, in respect to which the body S_2 is at rest. But the privileged space R_1 of Galileo, thus introduced, is a merely *factitious* cause, and not a thing that can be observed. It is therefore clear that Newton's mechanics does not really satisfy the requirement of causality in the case under consideration, but only apparently does so, since it makes the factitious cause R_1 responsible for the observable difference in the bodies S_1 and S_2 .

The only satisfactory answer must be that the physical system consisting of S_1 and S_2 reveals within itself no imaginable cause to which the differing behaviour of S_1 and S_2 can be referred. The cause must therefore lie *outside* this system. We have to take it that the general laws of motion, which in particular determine the shapes of S_1 and S_2, must be such that the mechanical behaviour of S_1 and S_2 is partly conditioned, in quite essential respects, by distant masses which we have not included in the system under consideration. These distant masses and their motions relative to S_1 and S_2 must then be regarded as the seat of the causes (which must be susceptible to observation) of the different behaviour of our two bodies S_1 and S_2. They take over the role of the factitious cause R_1. Of all imaginable spaces R_1, R_2, etc., in any kind of motion relatively to one another, there is none which we may look upon as privileged *a priori* without reviving the above-mentioned epistemological objection. *The laws of physics must be of such a nature that they apply to systems of reference in any hind of motion.* Along this road we arrive at an extension of the postulate of relativity.

In addition to this weighty argument from the theory of knowledge, there is a well-known physical fact which favours an extension of the theory of relativity. Let K be a Galilean system of reference, i.e. a system relatively to which (at least in the four-dimensional region under consideration) a mass, sufficiently distant from other

---

[2]Of course an answer may be satisfactory from the point of view of epistemology, and yet be unsound physically, if it is in conflict with other experiences.

masses, is moving with uniform motion in a straight line. Let K' be a second system of reference which is moving relatively to K in *uniformly accelerated* translation. Then, relatively to K', a mass sufficiently distant from other masses would have an accelerated motion such that its acceleration and direction of acceleration are independent of the material composition and physical state of the mass.

Does this permit an observer at rest relatively to K' to infer that he is on a "really" accelerated system of reference? The answer is in the negative;[3] for the above-mentioned relation of freely movable masses to K' may be interpreted equally well in the following way. The system of reference K' is unaccelerated, but the space-time territory in question is under the sway of a gravitational field, which generates the accelerated motion of the bodies relatively to K'.

This view is made possible for us by the teaching of experience as to the existence of a field of force, namely, the gravitational field, which possesses the remarkable property of imparting the same acceleration to all bodies.[4] The mechanical behaviour of bodies relatively to K' is the same as presents itself to experience in the case of systems which we are wont to regard as "stationary" or as "privileged." Therefore, from the physical standpoint, the assumption readily suggests itself that the systems K and K' may both with equal right be looked upon as "stationary," that is to say, they have an equal title as systems of reference for the physical description of phenomena.

It will be seen from these reflections that in pursuing the general theory of relativity we shall be led to a theory of gravitation, since we are able to "produce" a gravitational field merely by changing the system of coordinates. It will also be obvious that the principle of the constancy of the velocity of light *in vacuo* must be

---

[3]EDITOR'S NOTE: Even in 1916 Einstein seemed to have failed to appreciate the depth of Minkowski's explanation of why acceleration is absolute in spacetime (see EDITOR'S NOTE on p. 128) – "a somewhat curved worldline corresponds to a non-uniformly moving substantial point" (Minkowski, this volume, p. 110); so, acceleration is absolute because it reflects an *absolute* geometrical property of the worldline of an accelerating particle – its curvature, i.e., its *deformation*. Minkowski's explanation implies that the experimental fact (particles *resist* their acceleration), which itself *alone* proves that acceleration is absolute, has a profound physical meaning: the resistance a particle offers when accelerated, i.e., its inertia, is directly linked to the *deformation* of the worldline (rather worldtube) of the accelerating particle; unlike Poincaré (for whom what we now call spacetime was nothing more than a mathematical space), Minkowski regarded spacetime as representing a real four-dimensional world in which particles are real four-dimensional worldtubes (which implies that they resist their deformation). Obviously, Minkowski's explanation is for the case of flat spacetime, but it holds in curved spacetime as well. In curved spacetime a worldline can be naturally curved (due to the curvature of spacetime) or deformed (forcefully curved *by a force*). These two curvatures correspond to two types of acceleration in curved spacetime – relative (caused by geodesic deviation between two worldlines, which is a manifestation of the non-Euclidean geometry of spacetime) or absolute, which is a manifestation of the deformation of a *single* worldline and implies that, like in flat spacetime, an accelerated particle in curved spacetime (e.g., a particle on the Earth's surface) also resists its acceleration; therefore the weight of a particle, which was traditionally regarded as a gravitational force, is in fact an inertial force (Einstein wanted to demonstrate that inertial forces are gravitational, but, ironically, his general relativity showed that it is the opposite – even weight turned out to be an inertial force). So, as in the case of flat spacetime, the absolute acceleration in curved spacetime reflects an *absolute* geometrical property of the worldline of an accelerating particle – its curvature, i.e., its deformation. That is why an accelerating observer in curved spacetime can say that he *really* accelerates and he can indeed measure his acceleration with an accelerometer. Here Einstein introduces relative acceleration.

[4]Eötvos has proved experimentally that the gravitational field has this property in great accuracy.

modified, since we easily recognize that the path of a ray of light with respect to K' must in general be curvilinear, if with respect to K light is propagated in a straight line with a definite constant velocity.

§ 3. The Space-Time Continuum. Requirement of General covariance for the Equations Expressing General Laws of Nature

In classical mechanics, as well as in the special theory of relativity, the coordinates of space and time have a direct physical meaning. To say that a point-event has the X_1 coordinate x_1 means that the projection of the point-event on the axis of X_1, determined by rigid rods and in accordance with the rules of Euclidean geometry, is obtained by measuring off a given rod (the unit of length) x_1 times from the origin of coordinates along the axis of X_1. To say that a point-event has the X_4 coordinate $x_4 = t$, means that a standard clock, made to measure time in a definite unit period, and which is stationary relatively to the system of coordinates and practically coincident in space with the point-event,[5] will have measured off $x_4 = t$ periods at the occurrence of the event.

This view of space and time has always been in the minds of physicists, even if, as a rule, they have been unconscious of it. This is clear from the part which these concepts play in physical measurements; it must also have underlain the reader's reflections on the preceding paragraph (§ 2) for him to connect any meaning with what he there read. But we shall now show that we must put it aside and replace it by a more general view, in order to be able to carry through the postulate of general relativity, if the special theory of relativity applies to the special case of the absence of a gravitational field.

In a space which is free of gravitational fields we introduce a Galilean system of reference $K(x, y, z, t)$ and also a system of coordinates $K'(x', y', z', t')$ in uniform rotation relatively to K. Let the origins of both systems, as well as their axes of Z, permanently coincide. We shall show that for a space-time measurement in the system K' the above definition of the physical meaning of lengths and times cannot be maintained. For reasons of symmetry it is clear that a circle around the origin in the X, Y plane of K may at the same time be regarded as a circle in the X', Y' plane of K'. We suppose that the circumference and diameter of this circle have been measured with a unit measure infinitely small compared with the radius, and that we have the quotient of the two results. If this experiment were performed with a measuring-rod at rest relatively to the Galilean system K, the quotient would be π. With a measuring-rod at rest relatively to K', the quotient would be greater than π.[6] This is readily understood if we envisage the whole process of measuring

---

[5]We assume the possibility of verifying "simultaneity" for events immediately proximate in space, or—to speak more precisely—for immediate proximity or coincidence in space-time, without giving a definition of this fundamental concept.

[6]EDITOR'S NOTE: Einstein's assertion that the circumference of the rotating circle will be greater for a stationary observer is, unfortunately, incorrect. He erroneously assumed that the measuring-rod along the circumference contracts but the space (along the circumference) does not and therefore more measuring-rods will fit in the space along the circumference and the circumference will be longer (it will contain more measuring-rods) than when at rest (Einstein repeated this explanation

from the "stationary" system K, and take into consideration that the measuring-rod applied to the periphery undergoes a Lorentzian contraction, while the one applied along the radius does not. Hence Euclidean geometry does not apply to K'. The notion of coordinates defined above, which presupposes the validity of Euclidean geometry, therefore breaks down in relation to the system K'. So, too, we are unable to introduce a time corresponding to physical requirements in K', indicated by clocks

in his *Relativity: The Special and the General Theory. A Popular Exposition*. New publication in A. Einstein, *Relativity* (Minkowski Institute Press, Montreal 2018) p. 61). Regrettably, this Lorentzian view (that bodies contract but space itself does not) is a common misconception. This misconception can be immediately overcome when it is taken into account that the Lorentz transformations predict that the distance between two points in the space of a "stationary" observer, as measured by a moving observer, will be shorter than the distance between the same points measured by the "stationary" observer (at rest with respect to the points), *no matter whether these points are the end.points of a rod or just two points in space*.

The physical meaning of length contraction was made exceedingly clear by Minkowski in his ground-breaking lecture *Space and Time* delivered in 1908 (New translation (by V. Petkov) in: Hermann Minkowski, *Spacetime: Minkowski's Papers on Spacetime Physics*. Translated by Gregorie Dupuis-Mc Donald, Fritz Lewertoff and Vesselin Petkov. Edited by V. Petkov (Minkowski Institute Press, Montreal 2020), pp. 57-76.) – Minkowski showed that not only two observers in relative motion have different times but they also have different spaces (forming an angle) and these spaces intersect two parallel worldliness (representing either the end points of a rod or just two points in the space of one of the observers) under different angles; as a result the distance between the points will be different for the two observers.

In 1909 Ehrenfest arrived at the original formulation of the rotating disc problem (Ehrenfest considered a cylinder – P. Ehrenfest, Gleichförmige Rotation starrer Körper und Relativitätstheorie, *Physikalische Zeitschrift*, 1909, 10: 918) "on the basis of Minkowski's ideas" and correctly concluded that "the periphery of the cylinder has to show a contraction compared to its state of rest: $2\pi R' < 2\pi R$."

As there has been a lot of confusion in the literature about the rotating disc (due to involving irrelevant issues such as, for example, whether the disc would deform, which mixes real deformation effects caused by inertial forces that have noting to do with relativity with the relativistic issue of the geometry on the disc), the best way to understand this apparent paradox is by considering a geometric figure (as Einstein did) – a circle – and to follow explicitly Minkowski's ideas. Then the question of deformations (and even of rigidity) would not arise – as space itself contracts relativistically, the circumference of a rotating circle will be indisputably shorter than the circumference of a "stationary" circle. According to Minkowski the rotating circle is a twisted worldtube (a four-dimensional cylinder) in spacetime. The geometry of a rotating disc (based on that circle) is non-Euclidean for an imaginary observer at rest on the rotating disc. This is perhaps the most subtle and complicated part of this problem, because for that observer the surface of the disc will comprise a myriad of patches of the surface of the disc belonging to the spaces of a myriad of instntaneously comoving inertial observers (ICIOs) corresponding to a myriad of infinitesimal (relativistically contracted) segments of the disc's circumference measured by those ICIOs. However, the non-Euclidean geometry of the rotating disc is an artifact arising from the *description* of a four-dimensional object (the disc's worldtube) in our three-dimensional language (by using ICIOs) and does not represent anything real in the external world: (i) what is real in spacetime is the twisted worldtube of the rotating disc, not a three-dimensional disc (there are no three-dimensional objects in spacetime); (ii) the spaces of the ICIOs do not represent anything in spacetime either; they are just *descriptions* of spacetime in terms of our three-dimensional language.

That the non-Euclidean geometry of the rotating disc is an artifact is best seen by the fact that the disc rotates in *flat* (Euclidean) spacetime. The curvature (and therefore the geometry) of something (a line, surface or n-dimensional space) is an *absolute* frame-independent fact. Einstein's incorrect explanation might have been prompted by the behavior of unit measuring rods in the vicinity of a mass (along the radial direction towards the mass and along the tangential direction), discussed on p. 175. But these are two completely different cases since spacetime is non-Euclidean around the worldtube of the mass, whereas Ehrenfest's rotating disc (cylinder) is in flat spacetime.

at rest relatively to K'. To convince ourselves of this impossibility, let us imagine two clocks of identical constitution placed, one at the origin of coordinates, and the other at the circumference of the circle, and both envisaged from the "stationary" system K. By a familiar result of the special theory of relativity, the clock at the circumference—judged from K—goes more slowly than the other, because the former is in motion and the latter at rest. An observer at the common origin of coordinates, capable of observing the clock at the circumference by means of light, would therefore see it lagging behind the clock beside him. As he will not make up his mind to let the velocity of light along the path in question depend explicitly on the time, he will interpret his observations as showing that the clock at the circumference "really" goes more slowly than the clock at the origin. So he will be obliged to define time in such a way that the rate of a clock depends upon where the clock may be.

We therefore reach this result: In the general theory of relativity, space and time cannot be defined in such a way that differences of the spatial coordinates can be directly measured by the unit measuring-rod, or differences in the time coordinate by a standard clock.

The method hitherto employed for laying coordinates into the space-time continuum in a definite manner thus breaks down, and there seems to be no other way which would allow us to adapt systems of coordinates to the four-dimensional universe so that we might expect from their application a particularly simple formulation of the laws of nature. So there is nothing for it but to regard all imaginable systems of coordinates, on principle, as equally suitable for the description of nature. This comes to requiring that:

The general laws of nature are to be expressed by equations which hold good for all systems of coordinates, that is, are covariant with respect to any substitutions whatever (generally covariant).

It is clear that a physical theory which satisfies this postulate will also be suitable for the general postulate of relativity. For the sum of *all* substitutions in any case includes those which correspond to all relative motions of three-dimensional systems of coordinates. That this requirement of general covariance, which takes away from space and time the last remnant of physical objectivity, is a natural one, will be seen from the following reflection. All our space-time verifications invariably amount to a determination of space-time coincidences. If, for example, events consisted merely in the motion of material points, then ultimately nothing would be observable but the meetings of two or more of these points. Moreover, the results of our measurings are nothing but verifications of such meetings of the material points of our measuring instruments with other material points, coincidences between the hands of a clock and points on the clock dial, and observed point-events happening at the same place at the same time.

The introduction of a system of reference serves no other purpose than to facilitate the description of the totality of such coincidences. We allot to the universe four space-time variables x_1, x_2, x_3, x_4 in such a way that for every point-event there is a corresponding system of values of the variables $x_1 \ldots x_4$. To two coincident point-events there corresponds one system of values of the variables $x_1 \ldots x_4$, i.e. coincidence is characterized by the identity of the coordinates. If, in place of the variables $x_1 \ldots x_4$, we introduce functions of them, x'_1, x'_2, x'_3, x'_4, as a new system of coordinates, so that the systems of values are made to correspond to one another without ambiguity, the equality of all four coordinates in the new system will also serve as an expression for the space-time coincidence of the two point-events. As all our physical experience can be ultimately reduced to such coincidences, there is no immediate reason for preferring certain systems of coordinates to others, that is to say, we arrive at the requirement of general covariance.

§ 4. The Relation of the Four coordinates to Measurement in Space and Time

It is not my purpose in this discussion to represent the general theory of relativity as a system that is as simple and logical as possible, and with the minimum number of axioms; but my main object is to develop this theory in such a way that the reader will feel that the path we have entered upon is psychologically the natural one, and that the underlying assumptions will seem to have the highest possible degree of security. With this aim in view let it now be granted that:

For infinitely small four-dimensional regions the theory of relativity in the restricted sense is appropriate, if the coordinates are suitably chosen.

For this purpose we must choose the acceleration of the infinitely small ("local") system of coordinates so that no gravitational field occurs; this is possible for an infinitely small region. Let X_1, X_2, X_3, be the coordinates of space, and X_4 the appertaining coordinate of time measured in the appropriate unit.[7] If a rigid rod is imagined to be given as the unit measure, the coordinates, with a given orientation of the system of coordinates, have a direct physical meaning in the sense of the special theory of relativity. By the special theory of relativity the expression

$$ds^2 = -dX_1^2 - dX_2^2 - dX_3^2 + dX_4^2 \qquad (1)$$

then has a value which is independent of the orientation of the local system of coordinates, and is ascertainable by measurements of space and time. The magnitude of the linear element pertaining to points of the four-dimensional continuum in infinite proximity, we call ds. If the ds belonging to the element $X_1 \ldots X_4$ is positive, we follow Minkowski in calling it time-like; if it is negative, we call it space-like.

To the "linear element" in question, or to the two infinitely proximate point-events, there will also correspond definite differentials $dx_1 \ldots dx_4$ of the four-dimensional coordinates of any chosen system of reference. If this system, as well as the "local" system, is given for the region under consideration, the dX_ν will allow themselves to be represented here by definite linear homogeneous expressions of the dx_σ:

$$dX_\nu = \sum_\sigma a_{\nu\sigma} dx_\sigma \qquad (2)$$

---

[7] The unit of time is to be chosen so that the velocity of light in vacuo as measured in the "local" system of coordinates is to be equal to unity.

Inserting these expressions in (1), we obtain

$$ds^2 = \sum_{\tau\sigma} g_{\sigma\tau} dx_\sigma dx_\tau, \tag{3}$$

where the $g_{\sigma\tau}$ will be functions of the x_σ. These can no longer be dependent on the orientation and the state of motion of the "local" system of coordinates, for ds^2 is a quantity ascertainable by rod-clock measurement of point-events infinitely proximate in space-time, and defined independently of any particular choice of coordinates. The $g_{\sigma\tau}$ are to be chosen here so that $g_{\sigma\tau} = g_{\tau\sigma}$; the summation is to extend over all values of σ and τ, so that the sum consists of 4×4 terms, of which twelve are equal in pairs.

The case of the ordinary theory of relativity arises out of the case here considered, if it is possible, by reason of the particular relations of the $g_{\sigma\tau}$ in a finite region, to choose the system of reference in the finite region in such a way that the $g_{\sigma\tau}$ assume the constant values

$$\left. \begin{array}{cccc} -1 & 0 & 0 & 0 \\ 0 & -1 & 0 & 0 \\ 0 & 0 & -1 & 0 \\ 0 & 0 & 0 & +1 \end{array} \right\} \tag{4}$$

We shall find hereafter that the choice of such coordinates is, in general, not possible for a finite region.

From the considerations of § 2 and § 3 it follows that the quantities $g_{\tau\sigma}$ are to be regarded from the physical standpoint as the quantities which describe the gravitational field in relation to the chosen system of reference. For, if we now assume the special theory of relativity to apply to a certain four-dimensional region with the coordinates properly chosen, then the $g_{\sigma\tau}$ have the values given in (4). A free material point then moves, relatively to this system, with uniform motion in a straight line. Then if we introduce new space-time coordinates x_1, x_2, x_3, x_4 it by means of any substitution we choose, the $g^{\sigma\tau}$ in this new system will no longer be constants, but functions of space and time. At the same time the motion of the free material point will present itself in the new coordinates as a curvilinear non-uniform motion, and the law of this motion will be independent of the nature of the moving particle. We shall therefore interpret this motion as a motion under the influence of a gravitational field. We thus find the occurrence of a gravitational field connected with a space-time variability of the $g_{\sigma\tau}$. So, too, in the general case, when we are no longer able by a suitable choice of coordinates to apply the special theory of relativity to a finite region, we shall hold fast to the view that the $g_{\sigma\tau}$ describe the gravitational field.

Thus, according to the general theory of relativity, gravitation occupies an exceptional position with regard to other forces, particularly the electromagnetic forces, since the ten functions representing the gravitational field at the same time define the metrical properties of the space measured.

B. Mathematical Aids to the Formulation of Generally Covariant Equations

Having seen in the foregoing that the general postulate of relativity leads to the requirement that the equations of physics shall be covariant in the face of any substitution of the coordinates $x_1 \ldots x_4$, we have to consider how such generally covariant equations can be found. We now turn to this purely mathematical task, and we shall find that in its solution a fundamental rplee is played by the invariant ds given in equation (3), which, borrowing from Gauss's theory of surfaces, we have called the "linear element."

The fundamental idea of this general theory of covariants is the following: Let certain things ("tensors") be defined with respect to any system of coordinates by a number of functions of the coordinates, called the "components" of the tensor. There are then certain rules by which these components can be calculated for a new system of coordinates, if they are known for the original system of coordinates, and if the transformation connecting the two systems is known. The things hereafter called tensors are further characterized by the fact that the equations of transformation for their components are linear and homogeneous. Accordingly, all the components in the new system vanish, if they all vanish in the original system. If, therefore, a law of nature is expressed by equating all the components of a tensor to zero, it is generally covariant. By examining the laws of the formation of tensors, we acquire the means of formulating generally covariant laws.

§ 5. Contravariant and Covariant Four-vectors

Contravariant Four-vectors. The linear element is defined by the four "components" dx_ν, for which the law of transformation is expressed by the equation

$$dx'_\sigma = \sum_\nu \frac{\partial x'_\sigma}{\partial x_\nu} dx_\nu. \tag{5}$$

The $dx'\sigma$ are expressed as linear and homogeneous functions of the dx_ν. Hence we may look upon these coordinate differentials as the components of a "tensor" of the particular kind which we call a contravariant four-vector. Any thing which is defined relatively to the system of coordinates by four quantities A^ν, and which is transformed by the same law

$$A'^\sigma = \sum_\nu \frac{\partial x'_\sigma}{\partial x_\nu} A^\nu. \tag{5a}$$

we also call a contravariant four-vector. From (5a) it follows at once that the sums $A^\sigma \pm B^\sigma$ are also components of a four-vector, if A^σ and B^σ are such. Corresponding relations hold for all "tensors" subsequently to be introduced. (Rule for the addition and subtraction of tensors.)

Covariant Four-vectors. We call four quantities A_ν the components of a covariant four-vector, if for any arbitrary choice of the contravariant four-vector B^ν

$$\sum_\nu A_\nu B^\nu = \text{Invariant.} \tag{6}$$

The law of transformation of a covariant four-vector follows from this definition. For if we replace B^ν on the right-hand side of the equation

$$\sum_\sigma A'_\sigma B'^\sigma = \sum_\nu A_\nu B^\nu$$

by the expression resulting from the inversion of (5a),

$$\sum_\sigma \frac{\partial x_\nu}{\partial x'_\sigma} B'^\sigma,$$

we obtain

$$\sum_\sigma B'^\sigma \sum_\nu \frac{\partial x_\nu}{\partial x'_\sigma} A_\nu = \sum_\sigma B'^\sigma A'_\sigma.$$

Since this equation is true for arbitrary values of the B'^σ, it follows that the law of transformation is

$$A'_\sigma = \sum_\nu \frac{\partial x_\nu}{\partial x'_\sigma} A_\nu. \tag{7}$$

Note on a Simplified Way of Writing the Expressions. A glance at the equations of this paragraph shows that there is always a summation with respect to the indices which occur twice under a sign of summation (e.g. the index ν in (5)), and only with respect to indices which occur twice. It is therefore possible, without loss of clearness, to omit the sign of summation. In its place we introduce the convention: If an index occurs twice in one term of an expression, it is always to be summed unless the contrary is expressly stated.

The difference between covariant and contravariant four-vectors lies in the law of transformation ((7) or (5) respectively). Both forms are tensors in the sense of the general remark above. Therein lies their importance. Following Bicci and Levi-Civita, we denote the contravariant character by placing the index above, the covariant by placing it below.

§ 6. Tensors of the Second and Higher Ranks

Contravariant Tensors. If we form all the sixteen products $A^{\mu\nu}$ of the components A^μ and B^ν of two contravariant four-vectors

$$A^{\mu\nu} = A^\mu B^\nu \tag{8}$$

then by (8) and (5a) $A^{\mu\nu}$ satisfies the law of transformation

$$A'^{\sigma\tau} = \frac{\partial x'_\sigma}{\partial x_\mu} \frac{\partial x'_\tau}{\partial x_\nu} A^{\mu\nu}. \tag{9}$$

We call a thing which is described relatively to any system of reference by sixteen quantities, satisfying the law of transformation (9), a contravariant tensor of the second rank. Not every such tensor allows itself to be formed in accordance with

(8) from two four-vectors, but it is easily shown that any given sixteen $A^{\mu\nu}$ can be represented as the sums of the $A^\mu B^\nu$ of four appropriately selected pairs of four-vectors. Hence we can prove nearly all the laws which apply to the tensor of the second rank defined by (9) in the simplest manner by demonstrating them for the special tensors of the type (8).

Contravariant Tensors of Any Bank. It is clear that, on the lines of (8) and (9), contravariant tensors of the third and higher ranks may also be defined with 4^3 components, and so on. In the same way it follows from (8) and (9) that the contravariant four-vector may be taken in this sense as a contravariant tensor of the first rank.

Covariant Tensors. On the other hand, if we take the sixteen products $A_{\mu\nu}$ of two covariant four-vectors A_μ and B_ν,

$$A_{\mu\nu} = A_\mu B_\nu, \tag{10}$$

the law of transformation for these is

$$A'_{\sigma\tau} = \frac{\partial x_\mu}{\partial x'_\sigma} \frac{\partial x_\nu}{\partial x'_\tau} A_{\mu\nu}. \tag{11}$$

This law of transformation defines the covariant tensor of the second rank. All our previous remarks on contravariant tensors apply equally to covariant tensors.

NOTE. It is convenient to treat the scalar (or invariant) both as a contravariant and a covariant tensor of zero rank.

Mixed Tensors. We may also define a tensor of the second rank of the type

$$A_\mu^\nu = A_\mu B^\nu, \tag{12}$$

which is covariant with respect to the index μ, and contravariant with respect to the index ν. Its law of transformation is

$$A'^\tau_\sigma = \frac{\partial x'_\tau}{\partial x_\nu} \frac{\partial x_\mu}{\partial x'_\sigma} A_\mu^\nu. \tag{13}$$

Naturally there are mixed tensors with any number of indices of covariant character, and any number of indices of contravariant character. Covariant and contravariant tensors may be looked upon as special cases of mixed tensors.

Symmetrical Tensors. A contravariant, or a covariant tensor, of the second or higher rank is said to be symmetrical if two components, which are obtained the one from the other by the interchange of two indices, are equal. The tensor $A^{\mu\nu}$, or the tensor $A_{\mu\nu}$, is thus symmetrical if for any combination of the indices μ, ν,

$$A^{\mu\nu} = A^{\nu\mu}, \tag{14}$$

or respectively,

$$A_{\mu\nu} = A_{\nu\mu}. \tag{14a}$$

It has to be proved that the symmetry thus defined is a property which is independent of the system of reference. It follows in fact from (9), when (14) is taken into consideration, that

$$A'^{\sigma\tau} = \frac{\partial x'_\sigma}{\partial x_\mu} \frac{\partial x'_\tau}{\partial x_\nu} A^{\mu\nu} = \frac{\partial x'_\sigma}{\partial x_\mu} \frac{\partial x'_\tau}{\partial x_\nu} A^{\nu\mu} = \frac{\partial x'_\tau}{\partial x_\mu} \frac{\partial x'_\sigma}{\partial x_\nu} A^{\mu\nu} = A'^{\tau\sigma}.$$

The last equation but one depends upon the interchange of the summation indices μ, and ν, i.e. merely on a change of notation.

Antisymmetrical Tensors. A contravariant or a covariant tensor of the second, third, or fourth rank is said to be antisymmetrical if two components, which are obtained the one from the other by the interchange of two indices, are equal and of opposite sign. The tensor $A^{\mu\nu}$, or the tensor $A_{\mu\nu}$, is therefore antisymmetrical, if always

$$A^{\mu\nu} = -A^{\nu\mu}, \tag{15}$$

or respectively,

$$A_{\mu\nu} = -A_{\nu\mu}. \tag{15a}$$

Of the sixteen components $A^{\mu\nu}$ the four components $A^{\mu\mu}$ vanish; the rest are equal and of opposite sign in pairs, so that there are only six components numerically different (a six-vector). Similarly we see that the antisymmetrical tensor of the third rank $A^{\mu\nu\sigma}$ has only four numerically different components, while the antisymmetrical tensor $A^{\mu\nu\sigma\tau}$ has only one. There are no antisymmetrical tensors of higher rank than the fourth in a continuum of four dimensions.

§ 7. Multiplication of Tensors

Outer Multiplication of Tensors. We obtain from the components of a tensor of rank z and of a tensor of rank z' the components of a tensor of rank $z + z'$ by multiplying each component of the one tensor by each component of the other. Thus, for example, the tensors T arise out of the tensors A and B of different kinds,

$$T_{\mu\nu\sigma} = A_{\nu\mu} B_\sigma,$$
$$T^{\alpha\beta\gamma\delta} = A^{\alpha\beta} B^{\gamma\delta},$$
$$T^{\gamma\delta}_{\alpha\beta} = A_{\alpha\beta} B^{\gamma\delta}.$$

The proof of the tensor character of T is given directly by the representations (8), (10), (12), or by the laws of transformation (9), (11), (13). The equations (8), (10), (12) are themselves examples of outer multiplication of tensors of the first rank.

"Contraction" of a Mixed Tensor. From any mixed tensor we may form a tensor whose rank is less by two, by equating an index of covariant with one of contravariant character, and summing with respect to this index ("contraction"). Thus, for example, from the mixed tensor of the fourth rank $A^{\gamma\delta}_{\alpha\beta}$, we obtain the mixed tensor of the second rank,

$$A^\delta_\beta = A^{\gamma\delta}_{\alpha\beta} \left(= \sum_\alpha A^{\alpha\delta}_{\alpha\beta}\right)$$

and from this, by a second contraction, the tensor of zero rank,

$$A = A_\beta^\beta = A_{\alpha\beta}^{\alpha\beta}.$$

The proof that the result of contraction really possesses the tensor character is given either by the representation of a tensor according to the generalization of (12) in combination with (6), or by the generalization of (13).

Inner and Mixed Multiplication of Tensors. These consist in a combination of outer multiplication with contraction.

Examples. From the covariant tensor of the second rank $A_{\mu\nu}$ and the contravariant tensor of the first rank B^σ we form by outer multiplication the mixed tensor

$$D_{\mu\nu}^\sigma = A_{\mu\nu}B^\sigma.$$

On contraction with respect to the indices ν and σ, we obtain the covariant four-vector

$$D_\mu = D_{\mu\nu}^\nu = A_{\mu\nu}B^\nu.$$

This we call the inner product of the tensors $A_{\mu\nu}$ and B^σ. Analogously we form from the tensors $A_{\mu\nu}$ and $B^{\sigma\tau}$, by outer multiplication and double contraction, the inner product $A_{\mu\nu}B^{\mu\nu}$. By outer multiplication and one contraction, we obtain from $A_{\mu\nu}$ and $B^{\sigma\tau}$ the mixed tensor of the second rank $D_\mu^\tau = A_{\mu\nu}B^{\nu\tau}$. This operation may be aptly characterized as a mixed one, being "outer" with respect to the indices μ and τ, and "inner" with respect to the indices ν and σ.

We now prove a proposition which is often useful as evidence of tensor character. From what has just been explained, $A_{\mu\nu}B^{\mu\nu}$ is a scalar if $A_{\mu\nu}$ and $B^{\sigma\tau}$ are tensors. But we may also make the following assertion: If $A_{\mu\nu}B^{\mu\nu}$ is a scalar *for any choice of the tensor $B^{\mu\nu}$*, then $A_{\mu\nu}$ has tensor character. For, by hypothesis, for any substitution,

$$A_{\sigma\tau}'B'^{\sigma\tau} = A_{\mu\nu}B^{\mu\nu}.$$

But by an inversion of (9)

$$B^{\mu\nu} = \frac{\partial x_\mu}{\partial x_\sigma'}\frac{\partial x_\nu}{\partial x_\tau'} B'^{\sigma\tau}.$$

This, inserted in the above equation, gives

$$\left(A_{\sigma\tau}' - \frac{\partial x_\mu}{\partial x_\sigma'}\frac{\partial x_\nu}{\partial x_\tau'} A_{\mu\nu} \right) B'^{\sigma\tau} = 0.$$

This can only be satisfied for arbitrary values of $B'^{\sigma\tau}$ if the bracket vanishes. The result then follows by equation (11). This rule applies correspondingly to tensors of any rank and character, and the proof is analogous in all cases.

The rule may also be demonstrated in this form: If B^μ and C^ν are any vectors, and if, for all values of these, the inner product $A_{\mu\nu}B^\mu C^\nu$ is a scalar, then $A_{\mu\nu}$ is a covariant tensor. This latter proposition also holds good even if only the more special assertion is correct, that with any choice of the four-vector B^μ the inner product $A_{\mu\nu}B^\mu B^\nu$ is a scalar, if in addition it is known that $A_{\mu\nu}$ satisfies the condition of symmetry $A_{\mu\nu} = A_{\nu\mu}$. For by the method given above we prove the tensor character of $(A_{\mu\nu} + A_{\nu\mu})$, and from this the tensor character of $A_{\mu\nu}$ follows on account of symmetry. This also can be easily generalized to the case of covariant and contravariant tensors of any rank.

Finally, there follows from what has been proved, this law, which may also be generalized for any tensors: If for any choice of the four-vector B^ν the quantities $A_{\mu\nu}B^\nu$ form a tensor of the first rank, then $A_{\mu\nu}$ is a tensor of the second rank. For, if C^μ is any four-vector, then on account of the tensor character of $A_{\mu\nu}B^\nu$, the inner product $A_{\mu\nu}B^\nu C^\mu$ is a scalar for any choice of the two four-vectors B^ν and C^μ. From which the proposition follows.

§ 8. Some Aspects of the Fundamental Tensor

The Covariant Fundamental Tensor. In the invariant expression for the square of the linear element,

$$ds^2 = g_{\mu\nu}dx_\mu dx_\nu,$$

the part played by the dx is that of a contravariant vector which may be chosen at will. Since further, $g_{\mu\nu} = g_{\mu\mu}$, it follows from the considerations of the preceding paragraph that $g_{\mu\nu}$ is a covariant tensor of the second rank. We call it the "fundamental tensor." In what follows we deduce some properties of this tensor which, it is true, apply to any tensor of the second rank. But as the fundamental tensor plays a special part in our theory, which has its physical basis in the peculiar effects of gravitation, it so happens that the relations to be developed are of importance to us only in the case of the fundamental tensor.

The Contravariant Fundamental Tensor. If in the determinant formed by the elements $g_{\mu\nu}$, we take the co-factor of each of the $g_{\mu\nu}$ and divide it by the determinant $g = |g_{\mu\nu}|$, we obtain certain quantities $g^{\mu\nu}(= g^{\nu\mu})$ which, as we shall demonstrate, form a contravariant tensor.

By a known property of determinants

$$g_{\mu\sigma}g^{\nu\sigma} = \delta_\mu^\nu, \tag{16}$$

where the symbol δ_μ^ν denotes 1 or 0, according as $\mu = \nu$ or $\mu \neq \nu$.

Instead of the above expression for ds^2 we may thus write

$$g_{\mu\sigma}\delta_\nu^\sigma dx_\mu dx_\nu$$

or, by (16)

$$g_{\mu\sigma}g_{\nu\tau}g^{\sigma\tau}dx_\mu dx_\nu.$$

But, by the multiplication rules of the preceding paragraphs, the quantities

$$d\xi_\sigma = g_{\mu\sigma}dx_\mu$$

form a covariant four-vector, and in fact an arbitrary vector, since the dx_μ are arbitrary. By introducing this into our expression we obtain

$$ds^2 = g^{\sigma\tau}d\xi_\sigma d\xi_\tau.$$

Since this, with the arbitrary choice of the vector $d\xi_\sigma$, is a scalar, and $g^{\sigma\tau}$ by its definition is symmetrical in the indices σ and τ, it follows from the results of the preceding paragraph that $g^{\sigma\tau}$ is a contravariant tensor.

It further follows from (16) that δ_μ is also a tensor, which we may call the mixed fundamental tensor.

The Determinant of the Fundamental Tensor. By the rule for the multiplication of determinants

$$|g_{\mu\alpha}g^{\alpha\nu}| = |g_{\mu\alpha}|\,|g^{\alpha\nu}|\,.$$

On the other hand

$$|g_{\mu\alpha}g^{\alpha\nu}| = |\delta^\nu_\mu| = 1.$$

It therefore follows that

$$|g_{\mu\nu}|\,|g^{\mu\nu}| = 1. \tag{17}$$

The Volume Scalar. We seek first the law of transformation of the determinant $g = |g_{\mu\nu}|$. In accordance with (11)

$$g' = \left| \frac{\partial x_\mu}{\partial x'_\sigma} \frac{\partial x_\nu}{\partial x'_\tau} g_{\mu\nu} \right|\,.$$

Hence, by a double application of the rule for the multiplication of determinants, it follows that

$$g' = \left| \frac{\partial x_\mu}{\partial x'_\sigma} \right| \left| \frac{\partial x_\nu}{\partial x'_\tau} \right| |g_{\mu\nu}| = \left| \frac{\partial x_\mu}{\partial x'_\sigma} \right|^2 g,$$

or

$$\sqrt{g'} = \left| \frac{\partial x_\mu}{\partial x'_\sigma} \right| \sqrt{g}.$$

On the other hand, the law of transformation of the element of volume

$$d\tau = \int dx_1 dx_2 dx_3 dx_4$$

is, in accordance with the theorem of Jacobi,

$$d\tau' = \left| \frac{\partial x'_\sigma}{\partial x_\mu} \right| d\tau.$$

By multiplication of the last two equations, we obtain

$$\sqrt{g'}d\tau' = \sqrt{g}d\tau. \tag{18}$$

Instead of $\sqrt{g}$, we introduce in what follows the quantity $\sqrt{-g}$, which is always real on account of the hyperbolic character of the space-time continuum. The invariant $\sqrt{-g}d\tau$ is equal to the magnitude of the four-dimensional element of volume in the "local" system of reference, as measured with rigid rods and clocks in the sense of the special theory of relativity.

Note on the Character of the Space-time Continuum. Our assumption that the special theory of relativity can always be applied to an infinitely small region, implies that ds^2 can always be expressed in accordance with (1) by means of real quantities $dX_1 \ldots X_4$. If we denote by $d\tau_0$ the "natural" element of volume dX_1, X_2, dX_3, X_4, then

$$d\tau_0 = \sqrt{-g}\, d\tau. \tag{18a}$$

If $\sqrt{-g}$ were to vanish at a point of the four-dimensional continuum, it would mean that at this point an infinitely small "natural" volume would correspond to a finite volume in the coordinates. Let us assume that this is never the case. Then g cannot change sign. We will assume that, in the sense of the special theory of relativity, g always has a finite negative value. This is a hypothesis as to the physical nature of the continuum under consideration, and at the same time a convention as to the choice of coordinates.

But if $-g$ is always finite and positive, it is natural to settle the choice of coordinates *a posteriori* in such a way that this quantity is always equal to unity. We shall see later that by such a restriction of the choice of coordinates it is possible to achieve an important simplification of the laws of nature.

In place of (18), we then have simply $d\tau' = d\tau$, from which, in view of Jacobi's theorem, it follows that

$$\left| \frac{\partial x'_\sigma}{\partial x_\mu} \right| = 1. \tag{19}$$

Thus, with this choice of coordinates, only substitutions for which the determinant is unity are permissible.

But it would be erroneous to believe that this step indicates a partial abandonment of the general postulate of relativity. We do not ask "What are the laws of nature which are covariant in face of all substitutions for which the determinant is unity?" but our question is "What are the generally covariant laws of nature?" It is not until we have formulated these that we simplify their expression by a particular choice of the system of reference.

The Formation of New Tensors by Means of the Fundamental Tensor. Inner, outer, and mixed multiplication of a tensor by the fundamental tensor give tensors of different character and rank. For example,

$$A^\mu = g^{\mu\sigma} A_\sigma,$$
$$A = g_{\mu\nu} A^{\mu\nu}.$$

The following forms may be specially noted :

$$A^{\mu\nu} = g^{\mu\alpha} g^{\nu\beta} A_{\alpha\beta},$$
$$A_{\mu\nu} = g_{\mu\alpha} g_{\nu\beta} A^{\alpha\beta},$$

(the "complements" of covariant and contravariant tensors respectively), and

$$B_{\mu\nu} = g_{\mu\nu} g^{\alpha\beta} A_{\alpha\beta}.$$

We call $B_{\mu\nu}$ the reduced tensor associated with $A_{\mu\nu}$. Similarly,

$$B^{\mu\nu} = g^{\mu\nu} g_{\alpha\beta} A^{\alpha\beta}.$$

It may be noted that $g^{\mu\nu}$ is nothing more than the complement of $g_{\mu\nu}$, since

$$g^{\mu\alpha}g^{\nu\beta}g_{\alpha\beta} = g^{\mu\alpha}\delta^{\nu}_{\alpha} = g^{\mu\nu}.$$

§ 9. The Equation of the Geodetic Line (the Motion of a Particle)

As the linear element ds is defined independently of the system of coordinates, the line drawn between two points P and P' of the four-dimensional continuum in such a way that $\int ds$ is stationary—a geodetic line—has a meaning which also is independent of the choice of coordinates. Its equation is

$$\delta \int_{P}^{P'} ds = 0. \tag{20}$$

Carrying out the variation in the usual way, we obtain from this equation four differential equations which define the geodetic line; this operation will be inserted here for the sake of completeness. Let λ be a function of the coordinates x_{ν} and let this define a family of surfaces which intersect the required geodetic line as well as all the lines in immediate proximity to it which are drawn through the points P and P'. Any such line may then be supposed to be given by expressing its coordinates x_{ν} as functions of λ. Let the symbol δ indicate the transition from a point of the required geodetic to the point corresponding to the same λ on a neighbouring line. Then for (20) we may substitute

$$\left. \begin{aligned} \int_{\lambda_1}^{\lambda_2} \delta w \, d\lambda &= 0 \\ w^2 &= g_{\mu\nu} \frac{dx_\mu}{d\lambda} \frac{dx_\nu}{d\lambda} \end{aligned} \right\}. \tag{20a}$$

But since

$$\delta w = \frac{1}{w} \left\{ \frac{1}{2} \frac{\partial g_{\mu\nu}}{\partial x_\sigma} \frac{dx_\mu}{d\lambda} \frac{dx_\nu}{d\lambda} \delta x_\sigma + g_{\mu\nu} \frac{dx_\mu}{d\lambda} \delta \left(\frac{dx_\nu}{d\lambda} \right) \right\},$$

and

$$\delta \left(\frac{dx_\nu}{d\lambda} \right) = \frac{d \, \delta x_\nu}{d\lambda},$$

we obtain from (20a), after a partial integration,

$$\left. \begin{aligned} \int_{\lambda_1}^{\lambda_2} \kappa_\sigma \delta x_\sigma \, d\lambda &= 0, \\ \kappa_\sigma &= \frac{d}{d\lambda} \left\{ \frac{g_{\mu\nu}}{w} \frac{dx_\mu}{d\lambda} \right\} - \frac{1}{2w} \frac{\partial g_{\mu\nu}}{\partial x_\sigma} \frac{dx_\mu}{d\lambda} \frac{dx_\nu}{d\lambda} \end{aligned} \right\}. \tag{20b}$$

Since the values of δx_σ are arbitrary, it follows from this that

$$\kappa_\sigma = 0 \tag{20c}$$

are the equations of the geodetic line.

If ds does not vanish along the geodetic line we may choose the "length of the arc" s, measured along the geodetic line, for the parameter λ. Then $w = 1$, and in place of (20c) we obtain

$$g_{\mu\nu}\frac{d^2 x_\mu}{ds^2} + \frac{\partial g_{\mu\nu}}{\partial x_\sigma}\frac{dx_\sigma}{ds}\frac{dx_\mu}{ds} - \frac{1}{2}\frac{\partial g_{\mu\nu}}{\partial x_\sigma}\frac{dx_\mu}{ds}\frac{dx_\nu}{ds} = 0,$$

or, by a mere change of notation,

$$g_{\alpha\sigma}\frac{d^2 x_\alpha}{ds^2} + [\mu\nu,\,\sigma]\frac{dx_\mu}{ds}\frac{dx_\nu}{ds} = 0, \tag{20d}$$

where, following Christoffel, we have written

$$[\mu\nu,\,\sigma] = \frac{1}{2}\left(\frac{\partial g_{\mu\sigma}}{\partial x_\nu} + \frac{\partial g_{\nu\sigma}}{\partial x_\mu} - \frac{\partial g_{\mu\nu}}{\partial x_\sigma}\right), \tag{21}$$

Finally, if we multiply (20d) by $g^{\sigma\tau}$ (outer multiplication with respect to τ, inner with respect to σ), we obtain the equations of the geodetic line in the form

$$\frac{d^2 x_\tau}{ds^2} + \{\mu\nu,\,\tau\}\frac{dx_\mu}{ds}\frac{dx_\nu}{ds} = 0, \tag{22}$$

where, following Christoffel, we have set

$$\{\mu\nu,\,\tau\} = g^{\tau\alpha}\,[\mu\nu,\,\alpha]. \tag{23}$$

§10. The Formation of Tensors by Differentiation

With the help of the equation of the geodetic line we can now easily deduce the laws by which new tensors can be formed from old by differentiation. By this means we are able for the first time to formulate generally covariant differential equations. We reach this goal by repeated application of the following simple law:

If in our continuum a curve is given, the points of which are specified by the arcual distance s measured from a fixed point on the curve, and if, further, ϕ is an invariant function of space, then $d\phi/ds$ is also an invariant. The proof lies in this, that ds is an invariant as well as $d\phi$.

As

$$\frac{d\phi}{ds} = \frac{\partial\phi}{\partial x_\mu}\frac{dx_\mu}{ds}$$

therefore

$$\psi = \frac{\partial\phi}{\partial x_\mu}\frac{dx_\mu}{ds}$$

is also an invariant, and an invariant for all curves starting from a point of the continuum, that is, for any choice of the vector dx_μ. Hence it immediately follows that

$$A_\mu = \frac{\partial\phi}{\partial x_\mu} \tag{24}$$

is a covariant four-vector—the "gradient" of ϕ.

According to our rule, the differential quotient

$$\chi = \frac{d\psi}{ds}$$

taken on a curve, is similarly an invariant. Inserting the value of we ψ obtain in the first place

$$\chi = \frac{\partial^2 \phi}{\partial x_\mu \partial x_\nu} \frac{dx_\mu}{ds} \frac{dx_\nu}{ds} + \frac{\partial \phi}{\partial x_\mu} \frac{d^2 x_\mu}{ds^2}.$$

The existence of a tensor cannot be deduced from this forthwith. But if we may take the curve along which we have differentiated to be a geodetic, we obtain on substitution for $d^2 x_\nu / ds^2$ from (22),

$$\chi = \left(\frac{\partial^2 \phi}{\partial x_\mu \partial x_\nu} - \{\mu\nu, \tau\} \frac{\partial \phi}{\partial x_\tau} \right) \frac{dx_\mu}{ds} \frac{dx_\nu}{ds}.$$

Since we may interchange the order of the differentiations, and since by (23) and (21) $\{\mu\nu, \tau\}$ is symmetrical in μ and ν, it follows that the expression in brackets is symmetrical in μ and ν. Since a geodetic line can be drawn in any direction from a point of the continuum, and therefore dx_μ / ds is a four-vector with the ratio of its components arbitrary, it follows from the results of §7 that

$$A_{\mu\nu} = \frac{\partial^2 \phi}{\partial x_\mu \partial x_\nu} - \{\mu\nu, \tau\} \frac{\partial \phi}{\partial x_\tau} \tag{25}$$

is a covariant tensor of the second rank. We have therefore come to this result: from the covariant tensor of the first rank

$$A_\mu = \frac{\partial \phi}{\partial x_\mu}$$

we can, by differentiation, form a covariant tensor of the second rank

$$A_{\mu\nu} = \frac{\partial A_\mu}{\partial x_\nu} - \{\mu\nu, \tau\} A_\tau. \tag{26}$$

We call the tensor $A_{\mu\nu}$ the "extension" (covariant derivative) of the tensor A_μ. In the first place we can readily show that the operation leads to a tensor, even if the vector A_μ cannot be represented as a gradient. To see this, we first observe that

$$\psi \frac{\partial \phi}{\partial x_\mu}$$

is a covariant vector, if ψ and ϕ are scalars. The sum of four such terms

$$S_\mu = \psi^{(1)} \frac{\partial \phi^{(1)}}{\partial x_\mu} + \cdot + \cdot + \psi^{(4)} \frac{\partial \phi^{(4)}}{\partial x_\mu},$$

is also a covariant vector, if $\psi^{(1)}, \phi^{(1)} \ldots \psi^{(4)}\phi^{(4)}$ are scalars. But it is clear that any covariant vector can be represented in the form S_μ. For, if A_μ is a vector whose components are any given functions of the x_ν, we have only to put (in terms of the selected system of coordinates)

$$\psi^{(1)} = A_1, \qquad \phi^{(1)} = x_1,$$
$$\psi^{(2)} = A_2, \qquad \phi^{(2)} = x_2,$$
$$\psi^{(3)} = A_3, \qquad \phi^{(3)} = x_3,$$
$$\psi^{(4)} = A_4, \qquad \phi^{(4)} = x_4,$$

in order to ensure that S_μ shall be equal to A_μ.

Therefore, in order to demonstrate that $A_{\mu\nu}$ is a tensor if *any* covariant vector is inserted on the right-hand side for A_μ, we only need show that this is so for the vector S_μ. But for this latter purpose it is sufficient, as a glance at the right-hand side of (26) teaches us, to furnish the proof for the case

$$A_\mu = \psi \frac{\partial \phi}{\partial x_\mu}.$$

Now the right-hand side of (25) multiplied by

$$\psi \frac{\partial^2 \phi}{\partial x_\mu \partial x_\nu} - \{\mu\nu, \tau\} \psi \frac{\partial \phi}{\partial x_\tau}$$

is a tensor. Similarly

$$\frac{\partial \psi}{\partial x_\mu} \frac{\partial \phi}{\partial x_\nu}$$

being the outer product of two vectors, is a tensor. By addition, there follows the tensor character of

$$\frac{\partial}{\partial x_\nu} \left(\psi \frac{\partial \phi}{\partial x_\mu} \right) - \{\mu\nu, \tau\} \left(\psi \frac{\partial \phi}{\partial x_\tau} \right).$$

As a glance at (26) will show, this completes the demonstration for the vector

$$\psi \frac{\partial \phi}{\partial x_\mu}.$$

and consequently, from what has already been proved, for any vector A_μ.

By means of the extension of the vector, we may easily define the "extension" of a covariant tensor of any rank. This operation is a generalization of the extension of a vector. We restrict ourselves to the case of a tensor of the second rank, since this suffices to give a clear idea of the law of formation.

As has already been observed, any covariant tensor of the second rank can be represented[8] as the sum of tensors of the type $A_\mu B_\nu$. It will therefore be sufficient

---

[8] By outer multiplication of the veotor with arbitrary components $A_{11}, A_{12}, A_{13}, A_{14}$ by the vector

to deduce the expression for the extension of a tensor of this special type. By (26) the expressions

$$\frac{\partial A_\mu}{\partial x_\sigma} - \{\sigma\mu,\ \tau\}\, A_\tau,$$

$$\frac{\partial B_\nu}{\partial x_\sigma} - \{\sigma\nu,\ \tau\}\, B_\tau,$$

are tensors. On outer multiplication of the first by B_ν, and of the second by A_μ, we obtain in each case a tensor of the third rank. By adding these, we have the tensor of the third rank

$$A_{\mu\nu\sigma} = \frac{\partial A_{\mu\nu}}{\partial x_\sigma} - \{\sigma\mu,\ \tau\}\, A_{\tau\nu} - \{\sigma\nu,\ \tau\}\, A_{\mu\tau}, \tag{27}$$

where we have put $A_{\mu\nu} = A_\mu B_\nu$. As the right-hand side of (27) is linear and homogeneous in the $A_{\mu\nu}$ and their first derivatives, this law of formation leads to a tensor, not only in the case of a tensor of the type $A_\mu B_\nu$, but also in the case of a sum of such tensors, i.e. in the case of any covariant tensor of the second rank. We call $A_{\mu\nu\sigma}$ the extension of the tensor $A_{\mu\nu}$.

It is clear that (26) and (24) concern only special cases of extension (the extension of the tensors of rank one and zero respectively).

In general, all special laws of formation of tensors are included in (27) in combination with the multiplication of tensors.

§ 11. Some Cases of Special Importance

The Fundamental Tensor. We will first prove some lemmas which will be useful hereafter. By the rule for the differentiation of determinants

$$dg = g^{\mu\nu} g\, dg_{\mu\nu} = -g_{\mu\nu} g\, dg^{\mu\nu}. \tag{28}$$

The last member is obtained from the last but one, if we bear in mind that $g_{\mu\nu} g^{\mu'\nu} = \delta_\mu^{\mu'}$, so that $g_{\mu\nu} g^{\mu\nu} = 4$, and consequently

$$g_{\mu\nu} dg^{\mu\nu} + g^{\mu\nu} dg_{\mu\nu} = 0.$$

From (28), it follows that

$$\frac{1}{\sqrt{-g}}\frac{\partial\sqrt{-g}}{\partial x_\sigma} = \frac{1}{2}\frac{\partial\log(-g)}{\partial x_\sigma} = \frac{1}{2}g^{\mu\nu}\frac{\partial g_{\mu\nu}}{\partial x_\sigma} = \frac{1}{2}g_{\mu\nu}\frac{\partial g^{\mu\nu}}{\partial x_\sigma}. \tag{29}$$

with components $1, 0, 0, 0$, we produce a tensor with components

A_{11}	A_{12}	A_{13}	A_{14}
0	0	0	0
0	0	0	0
0	0	0	0

By the addition of four tensors of this type, we obtain the tensor $A_{\mu\nu}$ with any assigned components.

Further, from $g_{\mu\nu}g^{\mu\nu} = \delta^\mu_\mu$, it follows on differentiation that

$$\left.\begin{aligned}g_{\mu\sigma}dg^{\nu\sigma} &= -g^{\nu\sigma}dg_{\mu\sigma} \\ g_{\mu\sigma}\frac{\partial g^{\nu\sigma}}{\partial x_\lambda} &= -g^{\nu\sigma}\frac{\partial g_{\mu\sigma}}{\partial x_\lambda}\end{aligned}\right\} . \tag{30}$$

From these, by mixed multiplication by g aT and g v respectively, and a change of notation for the indices, we have

$$\left.\begin{aligned}dg^{\mu\nu} &= -g^{\mu\alpha}g^{\nu\beta}dg_{\alpha\beta} \\ \frac{\partial g^{\mu\nu}}{\partial x_\sigma} &= -g^{\mu\alpha}g^{\nu\beta}\frac{\partial g_{\alpha\beta}}{\partial x_\sigma}\end{aligned}\right\} \tag{31}$$

and

$$\left.\begin{aligned}dg_{\mu\nu} &= -g_{\mu\alpha}g_{\nu\beta}dg^{\alpha\beta} \\ \frac{\partial g_{\mu\nu}}{\partial x_\sigma} &= -g_{\mu\alpha}g_{\nu\beta}\frac{\partial g^{\alpha\beta}}{\partial x_\sigma}\end{aligned}\right\} . \tag{32}$$

The relation (31) admits of a transformation, of which we also have frequently to make use. From (21)

$$\frac{\partial g_{\alpha\beta}}{\partial x_\sigma} = [\alpha\sigma,\,\beta] + [\beta\sigma,\,\alpha] . \tag{33}$$

Inserting this in the second formula of (31), we obtain, in view of (23)

$$\frac{\partial g^{\mu\nu}}{\partial x_\sigma} = -g^{\mu\tau}\left\{\tau\sigma,\,\nu\right\} - g^{\nu\tau}\left\{\tau\sigma,\,\mu\right\} . \tag{34}$$

Substituting the right-hand side of (34) in (29), we have

$$\frac{1}{\sqrt{-g}}\frac{\partial\sqrt{-g}}{\partial x_\sigma} = \left\{\mu\sigma,\,\mu\right\} . \tag{29a}$$

The "Divergence" of a Contravariant Vector. If we take the inner product of (26) by the contravariant fundamental tensor $g^{\mu\nu}$, the right-hand side, after a transformation of the first term, assumes the form

$$\frac{\partial}{\partial x_\nu}\left(g^{\mu\nu}A_\mu\right) - A_\mu\frac{\partial g^{\mu\nu}}{\partial x_\nu} - \frac{1}{2}g^{\tau\alpha}\left(\frac{\partial g_{\mu\alpha}}{\partial x_\nu} + \frac{\partial g_{\nu\alpha}}{\partial x_\mu} - \frac{\partial g_{\mu\nu}}{\partial x_\alpha}\right)g^{\mu\nu}A_\nu .$$

In accordance with (31) and (29), the last term of this expression may be written

$$\frac{1}{2}\frac{\partial g^{\tau\nu}}{\partial x_\nu}A_\tau + \frac{1}{2}\frac{\partial g^{\tau\mu}}{\partial x_\mu}A_\tau + \frac{1}{\sqrt{-g}}\frac{\partial\sqrt{-g}}{\partial x_\alpha}g^{\mu\nu}A_\tau .$$

As the symbols of the indices of summation are immaterial, the first two terms of this expression cancel the second of the one above. If we then write $g^{\mu\nu}A_\mu = A^\nu$, so that A^ν like A_μ is an arbitrary vector, we finally obtain

$$\Phi = \frac{1}{\sqrt{-g}}\frac{\partial\sqrt{-g}}{\partial x_\nu}\left(\sqrt{-g}A^\nu\right) . \tag{35}$$

This scalar is the *divergence* of the contravariant vector A^ν.

The "Curl" of a Covariant Vector. The second term in (26) is symmetrical in the indices μ, and ν. Therefore $A_{\mu\mu} - A_{\nu\mu}$ is a particularly simply constructed antisymmetrical tensor. We obtain

$$B_{\mu\nu} = \frac{\partial A_\mu}{\partial x_\nu} - \frac{\partial A_\nu}{\partial x_\mu}. \tag{36}$$

Antisymmetrical Extension of a Six-vector. Applying (27) to an antisymmetrical tensor of the second rank $A_{\mu\nu}$, forming in addition the two equations which arise through cyclic permutations of the indices, and adding these three equations, we obtain the tensor of the third rank

$$B_{\mu\nu\sigma} = A_{\mu\nu\sigma} + A_{\nu\sigma\mu} + A_{\sigma\mu\nu} = \frac{\partial A_{\mu\nu}}{\partial x_\sigma} + \frac{\partial A_{\nu\sigma}}{\partial x_\mu} + \frac{\partial A_{\sigma\mu}}{\partial x_\nu}. \tag{37}$$

which it is easy to prove is antisymmetrical.

The Divergence of a Six-vector. Taking the mixed product of (27) by $g^{\mu\alpha}g^{\nu\beta}$, we also obtain a tensor. The first term on the right-hand side of (27) may be written in the form

$$\frac{\partial}{\partial x_\sigma}\left(g^{\mu\alpha}g^{\nu\beta}A_{\mu\nu}\right) - g^{\mu\alpha}\frac{\partial g^{\nu\beta}}{\partial x_\sigma}A_{\mu\nu} - g^{\nu\beta}\frac{\partial g^{\mu\alpha}}{\partial x_\sigma}A_{\mu\nu}.$$

If we write $A_\sigma^{\alpha\beta}$ for $g^{\mu\alpha}g^{\nu\beta}A_{\mu\nu\sigma}$ and $A^{\alpha\beta}$ for $g^{\mu\alpha}g^{\nu\beta}A_{\mu\nu}$, and in the transformed first term replace

$$\frac{\partial g^{\nu\beta}}{\partial x_\sigma} \quad \text{and} \quad \frac{\partial g^{\mu\alpha}}{\partial x_\sigma}$$

by their values as given by (34), there results from the right-hand side of (27) an expression consisting of seven terms, of which four cancel, and there remains

$$A_\sigma^{\alpha\beta} = \frac{\partial A^{\alpha\beta}}{\partial x_\sigma} + \{\sigma\gamma,\, \alpha\}\, A^{\gamma\beta} + \{\sigma\gamma,\, \beta\}\, A^{\alpha\gamma}. \tag{38}$$

This is the expression for the extension of a contravariant tensor of the second rank, and corresponding expressions for the extension of contravariant tensors of higher and lower rank may also be formed.

We note that in an analogous way we may also form the extension of a mixed tensor:

$$A_{\mu\sigma}^\alpha = \frac{\partial A_\mu^\alpha}{\partial x_\sigma} - \{\sigma\mu,\, \tau\}\, A_\tau^\alpha + \{\sigma\tau,\, \alpha\}\, A_\mu^\tau. \tag{39}$$

On contracting (38) with respect to the indices β and σ (inner multiplication by δ_β^σ), we obtain the vector

$$A^\alpha = \frac{\partial A^{\alpha\beta}}{\partial x_\beta} + \{\beta\gamma,\, \beta\}\, A^{\alpha\gamma} + \{\beta\gamma,\, \alpha\}\, A^{\gamma\beta}.$$

On account of the symmetry of $\{\beta\gamma,\, \alpha\}$ with respect to the indices β and γ, the third term on the right-hand side vanishes, if $A^{\alpha\beta}$ is, as we will assume, an antisymmetrical tensor. The second term allows itself to be transformed in accordance with (29a). Thus we obtain

$$A^\alpha = \frac{1}{\sqrt{-g}} \frac{\partial \left(\sqrt{-g} A^{\alpha\beta} \right)}{\partial x_\beta}. \tag{40}$$

This is the expression for the divergence of a contravariant six-vector.

   The Divergence of a Mixed Tensor of the Second Rank.   Contracting (39) with respect to the indices α and σ, and taking (29a) into consideration, we obtain

$$\sqrt{-g} A_\mu = \frac{\partial \left(\sqrt{-g} A_\mu^\sigma \right)}{\partial x_\sigma} - \{\sigma\mu, \tau\} \sqrt{-g} A_\tau^\sigma. \tag{41}$$

If we introduce the contravariant tensor $A^{\rho\sigma} = g^{\rho\tau} A_\tau^\sigma$ in the last term, it assumes the form

$$- \{\sigma\mu, \rho\} \sqrt{-g} A^{\rho\sigma}.$$

If, further, the tensor $A^{\rho\sigma}$ is symmetrical, this reduces to

$$-\frac{1}{2} \sqrt{-g} \frac{\partial g_{\rho\sigma}}{\partial x_\mu} A^{\rho\sigma}.$$

Had we introduced, instead of $A^{\rho\sigma}$, the covariant tensor $A_{\rho\sigma} = g_{\rho\alpha} g_{\sigma\beta} A^{\alpha\beta}$, which is also symmetrical, the last term, by virtue of (31), would assume the form

$$\frac{1}{2} \sqrt{-g} \frac{\partial g^{\rho\sigma}}{\partial x_\mu} A_{\rho\sigma}.$$

In the case of symmetry in question, (41) may therefore be replaced by the two forms

$$\sqrt{-g} A_\mu = \frac{\partial \left(\sqrt{-g} A_\mu^\sigma \right)}{\partial x_\sigma} - \frac{1}{2} \frac{\partial g_{\rho\sigma}}{\partial x_\mu} \sqrt{-g} A^{\rho\sigma}. \tag{41a}$$

$$\sqrt{-g} A_\mu = \frac{\partial \left(\sqrt{-g} A_\mu^\sigma \right)}{\partial x_\sigma} + \frac{1}{2} \frac{\partial g^{\rho\sigma}}{\partial x_\mu} \sqrt{-g} A_{\rho\sigma}. \tag{41b}$$

which we have to employ later on.

§ 12. The Riemann-Christoffel Tensor

   We now seek the tensor which can be obtained from the fundamental tensor *alone*, by differentiation. At first sight the solution seems obvious. We place the fundamental tensor of the $g_{\mu\nu}$ in (27) instead of any given tensor $A_{\mu\nu}$, and thus have a new tensor, namely, the extension of the fundamental tensor. But we easily convince ourselves that this extension vanishes identically. We reach our goal, however, in the following way. In (27) place

$$A_{\mu\nu} = \frac{\partial A_\mu}{\partial x_\nu} - \{\mu\nu, \rho\} A_\rho,$$

i.e. the extension of the four-vector A_μ. Then (with a somewhat different naming of the indices) we get the tensor of the third rank

$$A_{\mu\sigma\tau} = \frac{\partial^2 A_\mu}{\partial x_\sigma \partial x_\tau} - \{\mu\sigma, \rho\} \frac{\partial A_\rho}{\partial x_\tau} - \{\mu\tau, \rho\} \frac{\partial A_\rho}{\partial x_\sigma} - \{\sigma\tau, \rho\} \frac{\partial A_\mu}{\partial x_\rho}$$
$$+ \left[-\frac{\partial}{\partial x_\tau} \{\mu\sigma, \rho\} + \{\mu\tau, \alpha\}\{\alpha\sigma, \rho\} + \{\sigma\tau, \alpha\}\{\alpha\mu, \rho\} \right] A_\mu.$$

This expression suggests forming the tensor $A_{\mu\sigma\tau} - A_{\mu\tau\sigma}$. For, if we do so, the following terms of the expression for $A_{\mu\sigma\tau}$ cancel those of $A_{\mu\tau\sigma}$, the first, the fourth, and the member corresponding to the last term in square brackets; because all these are symmetrical in σ and τ. The same holds good for the sum of the second and third terms. Thus we obtain

$$A_{\mu\sigma\tau} - A_{\mu\tau\sigma} = B^\rho_{\mu\sigma\tau} A_\rho, \tag{42}$$

where

$$B^\rho_{\mu\sigma\tau} = -\frac{\partial}{\partial x_\tau}\{\mu\sigma, \rho\} + \frac{\partial}{\partial x_\sigma}\{\mu\tau, \rho\} - \{\mu\sigma, \alpha\}\{\alpha\tau, \rho\} + \{\mu\tau, \alpha\}\{\alpha\sigma, \rho\}. \tag{43}$$

The essential feature of the result is that on the right side of (42) the A_ρ occur alone, without their derivatives. From the tensor character of $A_{\mu\sigma\tau} - A_{\mu\tau\sigma}$ in conjunction with the fact that A_ρ is an arbitrary vector, it follows, by reason of § 7, that $B^\rho_{\mu\sigma\tau}$ is a tensor (the Riemann-Christoffel tensor).

The mathematical importance of this tensor is as follows: If the continuum is of such a nature that there is a coordinate system with reference to which the $g_{\mu\nu}$ are constants, then all the $B^\rho_{\mu\sigma\tau}$ vanish. If we choose any new system of coordinates in place of the original ones, the $g_{\mu\nu}$ referred thereto will not be constants, but in consequence of its tensor nature, the transformed components of $B^\rho_{\mu\sigma\tau}$ will still vanish in the new system. Thus the vanishing of the Riemann tensor is a necessary condition that, by an appropriate choice of the system of reference, the $g_{\mu\nu}$ may be constants. In our problem this corresponds to the case in which,[9] with a suitable choice of the system of reference, the special theory of relativity holds good for a *finite* region of the continuum.

Contracting (43) with respect to the indices τ and ρ we obtain the covariant tensor of second rank

$$G_{\mu\nu} = B^\rho_{\mu\nu\rho} = R_{\mu\nu} + S_{\mu\nu}$$

where

$$\left. \begin{array}{l} R_{\mu\nu} = -\dfrac{\partial}{\partial x_\alpha}\{\mu\nu, \alpha\} + \{\mu\alpha, \beta\}\{\nu\beta, \alpha\} \\[2ex] S_{\mu\nu} = \dfrac{\partial^2 \log\sqrt{-g}}{\partial x_\nu \partial x_\nu} - \{\mu\nu, \alpha\}\dfrac{\partial \log\sqrt{-g}}{\partial x_\alpha} \end{array} \right\} . \tag{44}$$

Note on the Choice of coordinates. It has already been observed in § 8, in connection with equation (18a), that the choice of coordinates may with advantage be

---

[9] The mathematicians have proved that this is also a *sufficient* condition.

made so that $\sqrt{-g} = 1$. A glance at the equations obtained in the last two sections shows that by such a choice the laws of formation of tensors undergo an important simplification. This applies particularly to $G_{\mu\nu}$, the tensor just developed, which plays a fundamental part in the theory to be set forth. For this specialization of the choice of coordinates brings about the vanishing of $S_{\mu\nu}$, so that the tensor $G_{\mu\nu}$ reduces to $R_{\mu\nu}$.

On this account I shall hereafter give all relations in the simplified form which this specialization of the choice of coordinates brings with it. It will then be an easy matter to revert to the *generally* covariant equations, if this seems desirable in a special case.

C. Theory of the Gravitational Field

§ 13. Equations of Motion of a Material Point in the Gravitational Field. Expression for the Field-components of Gravitation

A freely movable body not subjected to external forces moves, according to the special theory of relativity, in a straight line and uniformly. This is also the case, according to the general theory of relativity, for a part of four-dimensional space in which the system of coordinates K_0, may be, and is, so chosen that they have the special constant values given in (4).

If we consider precisely this movement from any chosen system of coordinates K_1, the body, observed from K_1, moves, according to the considerations in § 2, in a gravitational field. The law of motion with respect to K_1 results without difficulty from the following consideration. With respect to K_0 the law of motion corresponds to a four-dimensional straight line, i.e. to a geodetic line. Now since the geodetic line is defined independently of the system of reference, its equations will also be the equation of motion of the material point with respect to K_1. If we set

$$\Gamma^{\tau}_{\mu\nu} = - \{\mu\nu, \tau\} \tag{45}$$

the equation of the motion of the point with respect to K_1 becomes

$$\frac{d^2 x_{\tau}}{ds^2} = \Gamma^{\tau}_{\mu\nu} \frac{dx_{\mu}}{ds} \frac{dx_{\nu}}{ds}. \tag{46}$$

We now make the assumption, which readily suggests itself, that this covariant system of equations also defines the motion of the point in the gravitational field in the case when there is no system of reference K_0, with respect to which the special theory of relativity holds good in a finite region. We have all the more justification for this assumption as (46) contains only *first* derivatives of the $g_{\mu\nu}$, between which even in the special case of the existence of K_0, no relations subsist.[10]

If the $\Gamma^{\tau}_{\mu\nu}$ vanish, then the point moves uniformly in a straight line. These quantities therefore condition the deviation of the motion from uniformity. They are the components of the gravitational field.

---

[10]It is only between the second (and first) derivatives that, by § 12, the relations $B^{\rho}_{\mu\sigma\tau}$ subsist.

§ 14. The Field Equations of Gravitation in the Absence of Matter

We make a distinction hereafter between "gravitational field" and "matter" in this way, that we denote everything but the gravitational field as "matter." Our use of the word therefore includes not only matter in the ordinary sense, but the electromagnetic field as well.

Our next task is to find the field equations of gravitation in the absence of matter. Here we again apply the method employed in the preceding paragraph in formulating the equations of motion of the material point. A special case in which the required equations must in any case be satisfied is that of the special theory of relativity, in which the $g_{\mu\nu}$ have certain constant values. Let this be the case in a certain finite space in relation to a definite system of coordinates K_0. Relatively to this system all the components of the Riemann tensor $B^\rho_{\mu\sigma\tau}$, defined in (43), vanish. For the space under consideration they then vanish, also in any other system of coordinates.

Thus the required equations of the matter-free gravitational field must in any case be satisfied if all $B^\rho_{\mu\sigma\tau}$ vanish. But this condition goes too far. For it is clear that, e.g., the gravitational field generated by a material point in its environment certainly cannot be "transformed away" by any choice of the system of coordinates, i.e. it cannot be transformed to the case of constant $g_{\mu\nu}$.

This prompts us to require for the matter-free gravitational field that the symmetrical tensor $G_{\mu\nu}$, derived from the tensor $B^\rho_{\mu\sigma\tau}$. shall vanish. Thus we obtain ten equations for the ten quantities $g_{\mu\nu}$, which are satisfied in the special case of the vanishing of all $B^\rho_{\mu\sigma\tau}$. With the choice which we have made of a system of coordinates, and taking (44) into consideration, the equations for the matter-free field are

$$\left.\begin{array}{l} \dfrac{\partial \Gamma^\alpha_{\mu\nu}}{\partial} + \Gamma^\alpha_{\mu\beta}\Gamma^\beta_{\nu\alpha} \\[2mm] \sqrt{-g} = 1 \end{array}\right\}. \tag{47}$$

It must be pointed out that there is only a minimum of arbitrariness in the choice of these equations. For besides $G_{\mu\nu}$ there is no tensor of second rank which is formed from the $g_{\mu\nu}$ and its derivatives, contains no derivations higher than second, and is linear in these derivatives.[11]

These equations, which proceed, by the method of pure mathematics, from the requirement of the general theory of relativity, give us, in combination with the equations of motion (46), to a first approximation Newton's law of attraction, and to a second approximation the explanation of the motion of the perihelion of the planet Mercury discovered by Leverrier (as it remains after corrections for perturbation have been made). These facts must, in my opinion, be taken as a convincing proof of the correctness of the theory.

---

[11]Properly speaking, this can be affirmed only of the tensor

$$G_{\mu\nu} + \lambda g_{\mu\nu}(g^{\alpha\beta}G_{\alpha\beta}),$$

where λ is a constant. If, however, we set this tensor $= 0$, we come back again to the equations $G_{\mu\nu} = 0$.

§ 15. The Hamiltonian Function for the Gravitational Field. Laws of Momentum and Energy

To show that the field equations correspond to the laws of momentum and energy, it is most convenient to write them in the following Hamiltonian form:

$$\left.\begin{array}{l} \delta \int H d\tau = 0 \\ H = g^{\mu\nu}\Gamma^{\alpha}_{\mu\beta}\Gamma^{\beta}_{\nu\alpha} \\ \sqrt{-g} = 1 \end{array}\right\}, \tag{47a}$$

where, on the boundary of the finite four-dimensional region of integration which we have in view, the variations vanish.

We first have to show that the form (47a) is equivalent to the equations (47). For this purpose we regard H as a function of the $g^{\mu\nu}$ and the $g^{\mu\nu}_{\sigma}$ ($= \partial g^{\mu\nu}/\partial x_{\sigma}$).

Then in the first place

$$\delta H = \Gamma^{\alpha}_{\mu\beta}\Gamma^{\beta}_{\nu\alpha}\delta g^{\mu\nu} + 2g^{\mu\nu}\Gamma^{\alpha}_{\mu\beta}\delta\Gamma^{\beta}_{\nu\alpha}$$
$$= -\Gamma^{\alpha}_{\mu\beta}\Gamma^{\beta}_{\nu\alpha}\delta g^{\mu\nu} + 2\Gamma^{\alpha}_{\mu\beta}\delta(g^{\mu\nu}\Gamma^{\beta}_{\nu\alpha}).$$

But

$$\delta(g^{\mu\nu}\Gamma^{\beta}_{\nu\alpha}) - \frac{1}{2}\delta\left[g^{\mu\nu}g^{\beta\lambda}\left(\frac{\partial g_{\nu\lambda}}{\partial x_{\alpha}} + \frac{\partial g_{\alpha\lambda}}{\partial x_{\nu}} - \frac{\partial g_{\alpha\nu}}{\partial x_{\lambda}}\right)\right].$$

The terms arising from the last two terms in round brackets are of different sign, and result from each other (since the denomination of the summation indices is immaterial) through interchange of the indices μ and β. They cancel each other in the expression for δH, because they are multiplied by the quantity $\Gamma^{\alpha}_{\mu\beta}$, which is symmetrical with respect to the indices μ, and β. Thus there remains only the first term in round brackets to be considered, so that, taking (31) into account, we obtain

$$\delta H = -\Gamma^{\alpha}_{\mu\beta}\Gamma^{\beta}_{\nu\alpha}\delta g^{\mu\nu} + \Gamma^{\alpha}_{\mu\beta}\delta g^{\mu\beta}_{\alpha}.$$

Thus

$$\left.\begin{array}{l} \dfrac{\partial H}{\partial g^{\mu\nu}} = -\Gamma^{\alpha}_{\mu\beta}\Gamma^{\beta}_{\nu\alpha} \\ \dfrac{\partial H}{\partial g^{\mu\nu}_{\sigma}} = \Gamma^{\sigma}_{\mu\nu} \end{array}\right\}. \tag{48}$$

Carrying out the variation in (47a), we get in the first place

$$\frac{\partial}{\partial x_{\alpha}}\left(\frac{\partial H}{\partial g^{\mu\nu}_{\alpha}}\right) - \frac{\partial H}{\partial g^{\mu\nu}} = 0, \tag{47b}$$

which, on account of (48), agrees with (47), as was to be proved.

If we multiply (47b) by $g^{\mu\nu}_{\sigma}$, then because

$$\frac{\partial g^{\mu\nu}_{\sigma}}{\partial x_{\alpha}} = \frac{\partial g^{\mu\nu}_{\alpha}}{\partial x_{\sigma}}$$

and, consequently,

$$g_\sigma^{\mu\nu} \frac{\partial}{\partial x_\alpha} \left(\frac{\partial H}{\partial g_\alpha^{\mu\nu}} \right) = \frac{\partial}{\partial x_\alpha} \left(g_\sigma^{\mu\nu} \frac{\partial H}{\partial g_\alpha^{\mu\nu}} \right) - \frac{\partial H}{\partial g_\alpha^{\mu\nu}} \frac{\partial g_\alpha^{\mu\nu}}{\partial x_\sigma},$$

we obtain the equation

$$\frac{\partial}{\partial x_\alpha} \left(g_\sigma^{\mu\nu} \frac{\partial H}{\partial g_\alpha^{\mu\nu}} \right) - \frac{\partial H}{\partial x_\sigma} = 0$$

or[12]

$$\left. \begin{aligned} \frac{\partial t_\sigma^\alpha}{\partial x_\alpha} &= 0 \\ -2\kappa t_\sigma^\alpha &= g_\sigma^{\mu\nu} \frac{\partial H}{\partial g_\alpha^{\mu\nu}} - \delta_\sigma^\alpha H \end{aligned} \right\}, \tag{49}$$

where, on account of (48), the second equation of (47), and (34)

$$\kappa t_\sigma^\alpha = \tfrac{1}{2} \delta_\sigma^\alpha g^{\mu\nu} \Gamma_{\mu\beta}^\lambda \Gamma_{\nu\lambda}^\beta - g^{\mu\nu} \Gamma_{\mu\beta}^\alpha \Gamma_{\nu\sigma}^\beta \tag{50}$$

It is to be noticed that t_σ^α is not a tensor; on the other hand (49) applies to all systems of coordinates for which $\sqrt{-g} = 1$. This equation expresses the law of conservation of momentum and of energy for the gravitational field. Actually the integration of this equation over a three-dimensional volume V yields the four equations

$$\frac{d}{dx_4} \int t_\sigma^4 dV = \int \left(l t_\sigma^1 + m t_\sigma^2 + n t_\sigma^3 \right) dS, \tag{49a}$$

where l, m, n denote the direction-cosines of direction of the inward drawn normal at the element dS of the bounding surface (in the sense of Euclidean geometry). We recognize in this the expression of the laws of conservation in their usual form. The quantities t_σ^α we call the "energy components" of the gravitational field.

I will now give equations (47) in a third form, which is particularly useful for a vivid grasp of our subject. By multiplication of the field equations (47) by $g^{\nu\sigma}$ these are obtained in the "mixed" form. Note that

$$g^{\nu\sigma} \frac{\partial \Gamma_{\mu\nu}^\alpha}{\partial x_\alpha} = \frac{\partial}{\partial x_\alpha} \left(g^{\nu\sigma} \Gamma_{\mu\nu}^\alpha \right) - \frac{\partial g^{\nu\sigma}}{\partial x_\alpha} \Gamma_{\mu\nu}^\alpha,$$

which quantity, by reason of (34), is equal to

$$\frac{\partial}{\partial x_\alpha} \left(g^{\nu\sigma} \Gamma_{\mu\nu}^\alpha \right) - g^{\nu\beta} \Gamma_{\alpha\beta}^\sigma \Gamma_{\mu\nu}^\alpha - g^{\sigma\beta} \Gamma_{\beta\alpha}^\nu \Gamma_{\mu\nu}^\alpha,$$

or (with different symbols for the summation indices)

$$\frac{\partial}{\partial x_\alpha} \left(g^{\sigma\beta} \Gamma_{\mu\beta}^\alpha \right) - g^{\gamma\delta} \Gamma_{\gamma\beta}^\sigma \Gamma_{\delta\mu}^\beta - g^{\nu\sigma} \Gamma_{\mu\beta}^\alpha \Gamma_{\nu\alpha}^\beta.$$

---

[12]The reason for the introduction of the factor -2κ will be apparent later.

The third term of this expression cancels with the one arising from the second term of the field equations (47); using relation (50), the second term may be written

$$\kappa \left(t^{\sigma}_{\mu} - \tfrac{1}{2}\delta^{\sigma}_{\mu} t\right),$$

where $t = t^{\alpha}_{\alpha}$ Thus instead of equations (47) we obtain

$$\left.\begin{array}{c} \dfrac{\partial}{\partial x_{\alpha}}\left(g^{\sigma\beta}\Gamma^{\alpha}_{\mu\beta}\right) = -\kappa\left(t^{\sigma}_{\mu} - \tfrac{1}{2}\delta^{\sigma}_{\mu} t\right) \\[2ex] \sqrt{-g} = 1 \end{array}\right\}. \tag{51}$$

§ 16. The General Form of the Field Equations of Gravitation

The field equations for matter-free space formulated in § 15 are to be compared with the field equation

$$\nabla^2 \phi = 0$$

of Newton's theory. We require the equation corresponding to Poisson's equation

$$\nabla^2 \phi = 4\pi\kappa\rho,$$

where ρ denotes the density of matter.

The special theory of relativity has led to the conclusion that inert mass is nothing more or less than energy, which finds its complete mathematical expression in a symmetrical tensor of second rank, the energy-tensor. Thus in the general theory of relativity we must introduce a corresponding energy-tensor of matter T^{α}_{σ}, which, like the energy-components t^{α}_{σ} a [equations (49) and (50)] of the gravitational field, will have mixed character, but will pertain to a symmetrical covariant tensor.[13]

The system of equation (51) shows how this energy-tensor (corresponding to the density ρ in Poisson's equation) is to be introduced into the field equations of gravitation. For if we consider a complete system (e.g. the solar system), the total mass of the system, and therefore its total gravitating action as well, will depend on the total energy of the system, and therefore on the ponderable energy together with the gravitational energy. This will allow itself to be expressed by introducing into (51), in place of the energy-components of the gravitational field alone, the sums $t^{\sigma}_{\mu} + T^{\sigma}_{\mu}$ of the energy-components of matter and of gravitational field. Thus instead of (51) we obtain the tensor equation

$$\left.\begin{array}{c} \dfrac{\partial}{\partial x_{\alpha}}\left(g^{\sigma\beta}\Gamma^{\alpha}_{\mu\beta}\right) = -\kappa\left[\left(t^{\sigma}_{\mu} + T^{\sigma}_{\mu}\right) - \tfrac{1}{2}\delta^{\sigma}_{\mu}\left(t + T\right)\right] \\[2ex] \sqrt{-g} = 1 \end{array}\right\}, \tag{52}$$

where we have set $T = T^{\mu}_{\mu}$ (Laue's scalar). These are the required general field equations of gravitation in mixed form. Working back from these, we have in place

---

[13] $g_{\alpha\tau}T^{\alpha}_{\sigma} = T_{\sigma\tau}$ and $g^{\sigma\beta}T^{\alpha}_{\sigma} = T^{\alpha\beta}$ are to be symmetrical tensors.

of (47)

$$\frac{\partial}{\partial x_\alpha}\Gamma^\alpha_{\mu\nu} + \Gamma^\alpha_{\mu\beta}\Gamma^\beta_{\nu\alpha} = -\kappa\left(T_{\mu\nu} - \tfrac{1}{2}g_{\mu\nu}T\right),\\ \sqrt{-g} = 1 \Bigg\}. \tag{53}$$

It must be admitted that this introduction of the energy-tensor of matter is not justified by the relativity postulate alone. For this reason we have here deduced it from the requirement that the energy of the gravitational field shall act gravitatively in the same way as any other kind of energy. But the strongest reason for the choice of these equations lies in their consequence, that the equations of conservation of momentum and energy, corresponding exactly to equations (49) and (49a), hold good for the components of the total energy. This will be shown in § 17.

§ 17. The Laws of Conservation in the General Case

Equation (52) may readily be transformed so that the second term on the right-hand side vanishes. Contract (52) with respect to the indices μ and σ, and after multiplying the resulting equation by $\tfrac{1}{2}\delta^\sigma_\mu$, subtract it from equation (52). This gives

$$\frac{\partial}{\partial x_\alpha}\left(g^{\sigma\beta}\Gamma^\alpha_{\mu\beta} - \tfrac{1}{2}\delta^\sigma_\mu g^{\lambda\beta}\Gamma^\alpha_{\lambda\beta}\right) = -\kappa\left(t^\sigma_\mu + T^\sigma_\mu\right). \tag{52a}$$

On this equation we perform the operation $\partial/\partial x_\sigma$. We have

$$\frac{\partial^2}{\partial x_\alpha \partial x_\sigma}(g^{\sigma\beta}\Gamma^\alpha_{\mu\beta}) = -\frac{1}{2}\frac{\partial^2}{\partial x_\alpha \partial x_\sigma}\left[g^{\sigma\beta}g^{\alpha\lambda}\left(\frac{\partial g_{\mu\lambda}}{\partial x_\beta} + \frac{\partial g_{\beta\lambda}}{\partial x_\mu} - \frac{\partial g_{\mu\beta}}{\partial x_\lambda}\right)\right].$$

The first and third terms of the round brackets yield contributions which cancel one another, as may be seen by interchanging, in the contribution of the third term, the summation indices α and σ on the one hand, and β and λ on the other. The second term may be re-modelled by (31), so that we have

$$\frac{\partial^2}{\partial x_\alpha \partial x_\sigma}(g^{\sigma\beta}\Gamma^\alpha_{\mu\beta}) = \frac{1}{2}\frac{\partial^3 g^{\alpha\beta}}{\partial x_\alpha \partial x_\beta \partial x_\mu}. \tag{54}$$

The second term on the left-hand side of (52a) yields in the first place

$$-\frac{1}{2}\frac{\partial^2}{\partial x_\alpha \partial x_\mu}(g^{\lambda\beta}\Gamma^\alpha_{\lambda\beta})$$

or

$$\frac{1}{4}\frac{\partial^2}{\partial x_\alpha \partial x_\mu}\left[g^{\lambda\beta}g^{\alpha\delta}\left(\frac{\partial g_{\delta\lambda}}{\partial x_\beta} + \frac{\partial g_{\delta\beta}}{\partial x_\lambda} - \frac{\partial g_{\lambda\beta}}{\partial x_\delta}\right)\right].$$

With the choice of coordinates which we have made, the term deriving from the last term in round brackets disappears by reason of (29). The other two may be combined, and together, by (31), they give

$$-\frac{1}{2}\frac{\partial^3 g^{\alpha\beta}}{\partial x_\alpha \partial x_\beta \partial x_\mu},$$

so that in consideration of (54), we have the identity

$$\frac{\partial^2}{\partial x_\alpha \partial x_\sigma} \left(g^{\sigma\beta} \Gamma^\alpha_{\mu\beta} - \tfrac{1}{2} \delta^\sigma_\mu g^{\lambda\beta} \Gamma^\alpha_{\lambda\beta} \right) \equiv 0. \tag{55}$$

From (55) and (52a), it follows that

$$\frac{\partial \left(t^\sigma_\mu + T^\sigma_\mu \right)}{\partial x_\sigma} = 0. \tag{56}$$

Thus it results from our field equations of gravitation that the laws of conservation of momentum and energy are satisfied. This may be seen most easily from the consideration which leads to equation (49a); except that here, instead of the energy components t^σ_μ of the gravitational field, we have to introduce the totality of the energy components of matter and gravitational field.

§ 18. The Laws of Momentum and Energy for Matter as a Consequence of the Field Equations

Multiplying (53) by $\partial g^{\mu\nu}/\partial x_\sigma$ we obtain, by the method adopted in § 15, in view of the vanishing of

$$g_{\mu\nu} \frac{\partial g^{\mu\nu}}{\partial x_\sigma},$$

the equation

$$\frac{\partial t^\alpha_\sigma}{\partial x_\alpha} + \frac{1}{2} \frac{\partial g^{\mu\nu}}{\partial x_\sigma} T_{\mu\nu} = 0,$$

or, in view of (56),[14]

$$\frac{\partial T^\alpha_\sigma}{\partial x_\alpha} + \frac{1}{2} \frac{\partial g^{\mu\nu}}{\partial x_\sigma} T_{\mu\nu} = 0. \tag{57}$$

Comparison with (41b) shows that with the choice of system of coordinates which we have made, this equation predicates nothing more or less than the vanishing of divergence of the material energy-tensor. Physically, the occurrence of the second term on the left-hand side shows that laws of conservation of momentum and energy do not apply in the strict sense for matter alone, or else that they apply only when the g v are constant, i.e. when the field intensities of gravitation vanish. This second term is an expression for momentum, and for energy, as transferred per unit of volume and time from the gravitational field to matter. This is brought out still more clearly by re-writing (57) in the sense of (41) as

$$\frac{\partial T^\alpha_\sigma}{\partial x_\alpha} = -\Gamma^\beta_{\alpha\sigma} T^\alpha_\beta. \tag{57a}$$

---

[14]EDITOR'S NOTE: In view of (56)

$$\frac{\partial t^\alpha_\sigma}{\partial x_\alpha} = -\frac{\partial T^\alpha_\sigma}{\partial x_\alpha}.$$

and therefore the sign before the first term in (57) should be minus. Equation (57) above is displayed here as it appeared in the original 1916 German publication of Einstein's paper and the 1923 English translation.

The right side expresses the energetic effect of the gravitational field on matter.

Thus the field equations of gravitation contain four conditions which govern the course of material phenomena. They give the equations of material phenomena completely, if the latter is capable of being characterized by four differential equations independent of one another.[15]

D. "Material" Phenomena

The mathematical aids developed in part B enable us forthwith to generalize the physical laws of matter (hydrodynamics, Maxwell's electrodynamics), as they are formulated in the special theory of relativity, so that they will fit in with the general theory of relativity. When this is done, the general principle of relativity does not indeed afford us a further limitation of possibilities; but it makes us acquainted with the influence of the gravitational field on all processes, without our having to introduce any new hypothesis whatever.

Hence it comes about that it is not necessary to introduce definite assumptions as to the physical nature of matter (in the narrower sense). In particular it may remain an open question whether the theory of the electromagnetic field in conjunction with that of the gravitational field furnishes a sufficient basis for the theory of matter or not. The general postulate of relativity is unable on principle to tell us anything about this. It must remain to be seen, during the working out of the theory, whether electromagnetics and the doctrine of gravitation are able in collaboration to perform what the former by itself is unable to do.

§ 19. Euler's Equations for a Frictionless Adiabatic Fluid

Let p and ρ be two scalars, the former of which we call the "pressure," the latter the "density" of a fluid; and let an equation subsist between them. Let the contravariant symmetrical tensor

$$T^{\alpha\beta} = -g^{\alpha\beta}p + \rho\frac{dx_\alpha}{ds}\frac{dx_\beta}{ds} \tag{58}$$

be the contravariant energy-tensor of the fluid. To it belongs the covariant tensor

$$T_{\mu\nu} = -g_{\mu\nu}p + g_{\mu\alpha}\frac{dx_\alpha}{ds}g_{\mu\beta}\frac{dx_\beta}{ds}\rho, \tag{58a}$$

as well as the mixed tensor[16]

$$T_\sigma^\alpha = -\delta_\sigma^\alpha p + g_{\sigma\beta}\frac{dx_\beta}{ds}\frac{dx_\alpha}{ds}\rho. \tag{58b}$$

Inserting the right-hand side of (58b) in (57a), we obtain the Eulerian hydrodynamical equations of the general theory of relativity. They give, in theory, a complete solution

---

[15]On this question cf. D. Hilbert, Nachr. d. K. Gesellsch. d. Wiss. zu Göttingen, Math.-phys. Klasse, p. 3, 1915.

[16]For an observer using a system of reference in the sense of the special theory of relativity for an infinitely small region, and moving with it, the density of energy T_4^4 equals $\rho - p$. This gives the definition of ρ. Thus ρ is not constant for an incompressible fluid.

of the problem of motion, since the four equations (57a), together with the given equation between p and rho, and the equation

$$g_{\alpha\beta}\frac{dx_\alpha}{ds}\frac{dx_\beta}{ds} = 1,$$

are sufficient, $g_{\alpha\beta}$ being given, to define the six unknowns

$$p,\ \rho,\ \frac{dx_1}{ds},\ \frac{dx_2}{ds},\ \frac{dx_3}{ds},\ \frac{dx_4}{ds}.$$

If the $g_{\mu\nu}$ are also unknown, the equations (53) are brought in. These are eleven equations for defining the ten functions $g_{\mu\nu}$, so that these functions appear over-defined. We must remember, however, that the equations (57a) are already contained in the equations (53), so that the latter represent only seven independent equations. There is good reason for this lack of definition, in that the wide freedom of the choice of coordinates causes the problem to remain mathematically undefined to such a degree that three of the functions of space may be chosen at will.[17]

§ 20. Maxwell's Electromagnetic Field Equations for Free Space

Let ϕ_ν be the components of a covariant vector—the electromagnetic potential vector. From them we form, in accordance with (36), the components $F_{\rho\sigma}$ of the covariant six-vector of the electromagnetic field, in accordance with the system of equations

$$F_{\rho\sigma} = \frac{\partial\phi_\rho}{\partial x_\sigma} - \frac{\partial\phi_\sigma}{\partial x_\rho}. \tag{59}$$

It follows from (59) that the system of equations

$$\frac{\partial F_{\rho\sigma}}{\partial x_\tau} + \frac{\partial F_{\sigma\tau}}{\partial x_\rho} + \frac{\partial F_{\tau\rho}}{\partial x_\sigma} = 0 \tag{60}$$

is satisfied, its left side being, by (37), an antisymmetrical tensor of the third rank. System (60) thus contains essentially four equations which are written out as follows:

$$\left.\begin{aligned}
\frac{\partial F_{23}}{\partial x_4} + \frac{\partial F_{34}}{\partial x_2} + \frac{\partial F_{42}}{\partial x_3} &= 0 \\
\frac{\partial F_{34}}{\partial x_1} + \frac{\partial F_{41}}{\partial x_3} + \frac{\partial F_{13}}{\partial x_4} &= 0 \\
\frac{\partial F_{41}}{\partial x_2} + \frac{\partial F_{12}}{\partial x_4} + \frac{\partial F_{24}}{\partial x_1} &= 0 \\
\frac{\partial F_{12}}{\partial x_3} + \frac{\partial F_{23}}{\partial x_1} + \frac{\partial F_{31}}{\partial x_2} &= 0
\end{aligned}\right\}. \tag{60a}$$

[17]On the abandonment of the choice of coordinates with $g = -1$, there remain *four* functions of space with liberty of choice, corresponding to the four arbitrary functions at our disposal in the choice of coordinates.

This system corresponds to the second of Maxwell's systems of equations. We recognize this at once by setting

$$\left.\begin{array}{ll} F_{23} = H_x & F_{14} = E_x \\ F_{31} = H_y & F_{24} = E_y \\ F_{12} = H_z & F_{34} = E_z \end{array}\right\} . \tag{61}$$

Then in place of (60a) we may set, in the usual notation of three-dimensional vector analysis,

$$\left.\begin{array}{l} -\dfrac{\partial H}{\partial t} = \operatorname{curl} E \\ \operatorname{div} H = 0 \end{array}\right\} . \tag{60b}$$

We obtain Maxwell's first system by generalizing the form given by Minkowski. We introduce the contravariant six-vector associated with $F^{\alpha\beta}$

$$F^{\mu\nu} = g^{\mu\alpha} g^{\nu\beta} F_{\alpha\beta} \tag{62}$$

and also the contravariant vector J^{μ} of the density of the electric current. Then, taking (40) into consideration, the following equations will be invariant for any substitution whose invariant is unity (in agreement with the chosen coordinates):

$$\frac{\partial}{\partial x_{\nu}} F^{\mu\nu} = J^{\mu}. \tag{63}$$

Let

$$\left.\begin{array}{ll} F^{23} = H'_x & F^{14} = -E'_x \\ F^{31} = H'_y & F^{24} = -E'_y \\ F^{12} = H'_z & F^{34} = -E'_z \end{array}\right\} . \tag{64}$$

which quantities are equal to the quantities $H_x \ldots E_z$ in the special case of the restricted theory of relativity: and in addition

$$J^1 = j_x \quad J^2 = j_y \quad J^3 = j_z \quad J^4 = \rho,$$

we obtain in place of (63)

$$\left.\begin{array}{l} \dfrac{\partial E'}{\partial t} + j = \operatorname{curl} H' \\ \operatorname{div} E' = \rho \end{array}\right\} . \tag{63a}$$

The equations (60), (62), and (63) thus form the generalization of Maxwell's field equations for free space, with the convention which we have established with respect to the choice of coordinates.

The Energy-components of the Electromagnetic Field. We form the inner product

$$\kappa_{\sigma} = F_{\sigma\mu} J^{\mu}. \tag{}$$

By (61) its components, written in the three-dimensional manner, are

$$\left.\begin{aligned}\kappa_1 &= \rho E_x + [j \cdot H]_x \\ &\cdots\cdots\cdots\cdots\cdots \\ &\cdots\cdots\cdots\cdots\cdots \\ \kappa_4 &= -(jE)\end{aligned}\right\}. \tag{65a}$$

κ_σ is a covariant vector the components of which are equal to the negative momentum, or, respectively, the energy, which is transferred from the electric masses to the electromagnetic field per unit of time and volume. If the electric masses are free, that is, under the sole influence of the electromagnetic field, the covariant vector κ_σ will vanish.

To obtain the energy-components T_σ^ν of the electromagnetic field, we need only give to equation $\kappa_\sigma = 0$ the form of equation (57). From (63) and (65) we have in the first place

$$\kappa_\sigma = F_{\sigma\mu}\frac{\partial F^{\mu\nu}}{\partial x_\nu} = \frac{\partial}{\partial x_\nu}(F_{\sigma\mu}F^{\mu\nu}) - F^{\mu\nu}\frac{\partial F_{\sigma\mu}}{\partial x_\nu}.$$

The second term of the right-hand side, by reason of (60), permits the transformation

$$F^{\mu\nu}\frac{\partial F_{\sigma\mu}}{\partial x_\nu} = -\frac{1}{2}F^{\mu\nu}\frac{\partial F_{\mu\nu}}{\partial x_\sigma} = -\frac{1}{2}g^{\mu\alpha}g^{\nu\beta}F_{\alpha\beta}\frac{\partial F_{\mu\nu}}{\partial x_\sigma},$$

which latter expression may, for reasons of symmetry, also be written in the form

$$-\frac{1}{4}\left[g^{\mu\alpha}g^{\nu\beta}F_{\alpha\beta}\frac{\partial F_{\mu\nu}}{\partial x_\sigma} + g^{\mu\alpha}g^{\nu\beta}\frac{\partial F_{\alpha\beta}}{\partial x_\sigma}F_{\mu\nu}\right].$$

But for this we may set

$$-\frac{1}{4}\frac{\partial}{\partial x_\sigma}\left(g^{\mu\alpha}g^{\nu\beta}F_{\alpha\beta}F_{\mu\nu}\right) + \frac{1}{4}F_{\alpha\beta}F_{\mu\nu}\frac{\partial}{\partial x_\sigma}\left(g^{\mu\alpha}g^{\nu\beta}\right).$$

The first of these terms is written more briefly

$$-\frac{1}{4}\frac{\partial}{\partial x_\sigma}\left(F^{\mu\nu}F_{\mu\nu}\right);$$

the second, after the differentiation is carried out, and after some reduction, results in

$$-\frac{1}{2}F^{\mu\tau}F_{\mu\nu}g^{\nu\rho}\frac{\partial g^{\sigma\tau}}{\partial x_\sigma}.$$

Taking all three terms together we obtain the relation

$$\kappa_\sigma = \frac{\partial T_\sigma^\nu}{\partial x_\nu} - \frac{1}{2}g^{\tau\mu}\frac{\partial g_{\mu\nu}}{\partial x_\sigma}T_\tau^\nu, \tag{66}$$

where

$$T_\sigma^\nu = -F_{\sigma\alpha}F^{\nu\alpha} + \frac{1}{4}\delta_\sigma^\nu F_{\alpha\beta}F^{\alpha\beta}.$$

Equation (66), if k<, vanishes, is, on account of (30), equivalent to (57) or (57a) respectively. Therefore the T_σ^ν are the energy-components of the electromagnetic field. With the help of (61) and (64), it is easy to show that these energy-components of the electromagnetic field in the case of the special theory of relativity give the well-known Maxwell-Poynting expressions.

We have now deduced the general laws which are satisfied by the gravitational field and matter, by consistently using a system of coordinates for which $\sqrt{-g} = 1$. We have thereby achieved a considerable simplification of formulae and calculations, without failing to comply with the requirement of general covariance; for we have drawn our equations from generally covariant equations by specializing the system of coordinates.

Still the question is not without a formal interest, whether with a correspondingly generalized definition of the energy-components of gravitational field and matter, even without specializing the system of coordinates, it is possible to formulate laws of conservation in the form of equation (56), and field equations of gravitation of the same nature as (52) or (52a), in such a manner that on the left we have a divergence (in the ordinary sense), and on the right the sum of the energy-components of matter and gravitation. I have found that in both cases this is actually so. But I do not think that the communication of my somewhat extensive reflections on this subject would be worth while, because after all they do not give us anything that is materially new.

E

§ 21. Newton's Theory as a First Approximation

As has already been mentioned more than once, the special theory of relativity as a special case of the general theory is characterized by the $g_{\mu\nu}$ having the constant values (4). From what has already been said, this means complete neglect of the effects of gravitation. We arrive at a closer approximation to reality by considering the case where the $g_{\mu\nu}$ differ from the values of (4) by quantities which are small compared with 1, and neglecting small quantities of second and higher order. (First point of view of approximation.)

It is further to be assumed that in the space-time territory under consideration the $g_{\mu\nu}$ at spatial infinity, with a suitable choice of coordinates, tend toward the values (4); i.e. we are considering gravitational fields which may be regarded as generated exclusively by matter in the finite region.

It might be thought that these approximations must lead us to Newton's theory. But to that end we still need to approximate the fundamental equations from a second point of view. We give our attention to the motion of a material point in accordance with the equations (16). In the case of the special theory of relativity the components

$$\frac{dx_1}{ds}, \; \frac{dx_2}{ds}, \; \frac{dx_3}{ds}$$

may take on any values. This signifies that any velocity

$$v = \sqrt{\left(\frac{dx_1}{dx_4}\right)^2 + \left(\frac{dx_2}{dx_4}\right)^2 + \left(\frac{dx_3}{dx_4}\right)^2}$$

may occur, which is less than the velocity of light *in vacuo*. If we restrict ourselves to the case which almost exclusively offers itself to our experience, of v being small as compared with the velocity of light, this denotes that the components

$$\frac{dx_1}{ds}, \frac{dx_2}{ds}, \frac{dx_3}{ds}$$

are to be treated as small quantities, while dx_4/ds, to the second order of small quantities, is equal to one. (Second point of view of approximation.)

Now we remark that from the first point of view of approximation the magnitudes $\Gamma^\tau_{\mu\nu}$ are all small magnitudes of at least the first order. A glance at (46) thus shows that in this equation, from the second point of view of approximation, we have to consider only terms for which $\mu = \nu = 4$. Restricting ourselves to terms of lowest order we first obtain in place of (46) the equations

$$\frac{d^2 x_\tau}{dt^2} = \Gamma^\tau_{44}$$

where we have set $ds = dx_4 = dt$; or with restriction to terms which from the first point of view of approximation are of first order:

$$\frac{d^2 x_\tau}{dt^2} = [44, \tau] \quad (\tau = 1, 2, 3)$$

$$\frac{d^2 x_4}{dt^2} = -[44, 4].$$

If in addition we suppose the gravitational field to be a quasistatic field, by confining ourselves to the case where the motion of the matter generating the gravitational field is but slow (in comparison with the velocity of the propagation of light), we may neglect on the right-hand side differentiations with respect to the time in comparison with those with respect to the space coordinates, so that we have

$$\frac{d^2 x_\tau}{dt^2} = -\frac{1}{2}\frac{\partial g_{44}}{\partial x_\tau} \quad (\tau = 1, 2, 3) \tag{67}$$

This is the equation of motion of the material point according to Newton's theory, in which $\frac{1}{2}g_{44}$ plays the part of the gravitational potential. What is remarkable in this result is that the component g_{44} of the fundamental tensor alone defines, to a first approximation, the motion of the material point.

We now turn to the field equations (53). Here we have to take into consideration that the energy-tensor of "matter" is almost exclusively defined by the density of matter in the narrower sense, i.e. by the second term of the right-hand side of (58) [or, respectively, (58a) or (58b)]. If we form the approximation in question, all the components vanish with the one exception of

$$T_{44} = \rho = T.$$

On the left-hand side of (53) the second term is a small quantity of second order; the first yields, to the approximation in question,

$$\frac{\partial}{\partial x_1}\,[\mu\nu,\,1] + \frac{\partial}{\partial x_2}\,[\mu\nu,\,2] + \frac{\partial}{\partial x_3}\,[\mu\nu,\,3] - \frac{\partial}{\partial x_4}\,[\mu\nu,\,4]\,.$$

For $\mu = \nu = 4$, this gives, with the omission of terms differentiated with respect to time,

$$-\frac{1}{2}\left(\frac{\partial^2 g_{44}}{\partial x_1^2} + \frac{\partial^2 g_{44}}{\partial x_2^2} + \frac{\partial^2 g_{44}}{\partial x_3^2}\right) = -\frac{1}{2}\nabla^2 g_{44}.$$

The last of equations (53) thus yields

$$\nabla^2 g_{44} = \kappa\rho. \tag{68}$$

The equations (67) and (68) together are equivalent to Newton's law of gravitation. By (67) and (68) the expression for the gravitational potential becomes

$$-\frac{\kappa}{8\pi}\int\frac{\rho\,d\tau}{r}, \tag{68a}$$

while Newton's theory, with the unit of time which we have chosen, gives

$$-\frac{K}{c^2}\int\frac{\rho\,d\tau}{r},$$

in which K denotes the constant 6.7×10^{-8}, usually called the constant of gravitation. By comparison we obtain

$$\kappa = \frac{8\pi K}{c^2} = 1.87 \times 10^{-27}. \tag{69}$$

§ 22. Behaviour of Rods and Clocks in the Static Gravitational Field. Bending of Light-Rays. Motion of the Perihelion of a Planetary Orbit

To arrive at Newton's theory as a first approximation we had to calculate only one component, g_{44} of the ten $g_{\mu\nu}$ of the gravitational field, since this component alone enters into the first approximation, (67), of the equation for the motion of the material point in the gravitational field. From this, however, it is already apparent that other components of the $g_{\mu\nu}$ must differ from the values given in (4) by small quantities of the first order. This is required by the condition $g = -1$.

For a field-producing point mass at the origin of coordinates, we obtain, to the first approximation, the radially symmetrical solution

$$
\left.\begin{aligned}
g_{\rho\sigma} &= -\delta_{\rho\sigma} - \alpha \frac{x_\rho x_\sigma}{r^3} \quad (\rho, \sigma = 1, 2, 3) \\
g_{\rho 4} &= g_{4\rho} = 0 \qquad\qquad (\rho = 1, 2, 3) \\
g_{44} &= 1 - \frac{\alpha}{r}
\end{aligned}\right\}. \tag{70}
$$

where $\delta_{\rho\sigma}$ is 1 or 0, respectively, accordingly as $\rho = \sigma$ or $\rho \neq \sigma$, and r is the quantity $+\sqrt{x_1^2 + x_2^2 + x_3^2}$. On account of (68a)

$$
\alpha = \frac{\kappa M}{4\pi} \tag{70a}
$$

if M denotes the field-producing mass. It is easy to verify that the field equations (outside the mass) are satisfied to the first order of small quantities.

We now examine the influence exerted by the field of the mass M upon the metrical properties of space. The relation

$$
ds^2 = g_{\mu\nu} dx_\mu dx_\nu
$$

always holds between the "locally" (§ 4) measured lengths and times ds on the one hand, and the differences of coordinates dx_ν on the other hand.

For a unit-measure of length laid "parallel" to the axis of x, for example, we should have to set $ds^2 = -1$; $dx_2 = dx_3 = dx_4 = 0$. Therefore $-1 = g_{11} dx_1^2$. If, in addition, the unit-measure lies on the axis of x, the first of equations (70) gives

$$
g_{11} = -\left(1 + \frac{\alpha}{r}\right).
$$

From these two relations it follows that, correct to a first order of small quantities,

$$
dx = 1 - \frac{\alpha}{2r} \tag{71}
$$

The unit measuring-rod thus appears a little shortened in relation to the system of coordinates by the presence of the gravitational field, if the rod is laid along a radius.

In an analogous manner we obtain the length of coordinates in tangential direction if, for example, we set

$$
ds^2 = -1; \quad dx_1 = dx_3 = dx_4 = 0; \quad x_1 = r, \ x_2 = x_3 = 0.
$$

The result is

$$
-1 = g_{22} dx_2^2 = -dx_2^2. \tag{71a}
$$

With the tangential position, therefore, the gravitational field of the point of mass has no influence on the length of a rod.

Thus Euclidean geometry does not hold even to a first approximation in the gravitational field, if we wish to take one and the same rod, independently of its place and orientation, as a realization of the same interval; although, to be sure, a glance at (70a) and (69) shows that the deviations to be expected are much too slight to be noticeable in measurements of the earth's surface.

Further, let us examine the rate of a unit clock, which is arranged to be at rest in a static gravitational field. Here we have for a clock period $ds = 1$; $dx_1 = dx_2 = dx_3 = 0$. Therefore

$$1 = g_{44}\, dx_4^2;$$

$$dx_4 = \frac{1}{\sqrt{g_{44}}} = \frac{1}{\sqrt{1 + (g_{44} - 1)}} = 1 - \frac{1}{2}(g_{44} - 1)$$

or

$$dx_4 = 1 + \frac{\kappa}{8\pi} \int \frac{\rho d\tau}{r}. \tag{72}$$

Thus the clock goes more slowly if set up in the neighbourhood of ponderable masses. From this it follows that the spectral lines of light reaching us from the surface of large stars must appear displaced towards the red end of the spectrum.[18]

We now examine the course of light-rays in the static gravitational field. By the special theory of relativity the velocity of light is given by the equation

$$-dx_1^2 - dx_2^2 - dx_3^2 + dx_4^2 = 0$$

and therefore by the general theory of relativity by the equation

$$ds^2 = g_{\mu\nu}dx_\mu dx_\nu = 0. \tag{73}$$

If the direction, i.e. the ratio $dx_1 : dx_2 : dx_3$ is given, equation (73) gives the quantities

$$\frac{dx_1}{dx_4}, \frac{dx_2}{dx_4}, \frac{dx_3}{dx_4}$$

and accordingly the velocity

$$\sqrt{\left(\frac{dx_1}{dx_4}\right)^2 + \left(\frac{dx_2}{dx_4}\right)^2 + \left(\frac{dx_3}{dx_4}\right)^2} = \gamma$$

defined in the sense of Euclidean geometry. We easily recognize that the course of the light-rays must be bent with regard to the system of coordinates, if the $g_{\mu\nu}$ are not constant. If n is a direction perpendicular to the propagation of light, the Huyghens principle shows that the light-ray, envisaged in the plane (γ, n), has the curvature $-\partial\gamma/\partial n$.

We examine the curvature undergone by a ray of light passing by a mass M at the distance Δ. If we choose the system of coordinates in agreement with the accompanying diagram, the total bending of the ray (calculated positively if concave towards the origin) is given in sufficient approximation by

$$B = \int_{-\infty}^{+\infty} \frac{\partial\gamma}{\partial x_1}\, dx_2,$$

---

[18] According to E. Freundlich, spectroscopical observations on fixed stars of certain types indicate the existence of an effect of this kind, but a crucial test of this consequence has not yet been made.

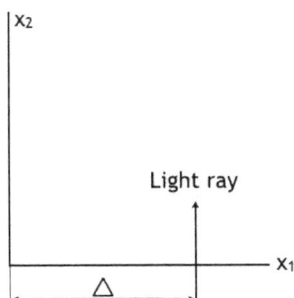

while (78) and (70) give

$$\gamma = \sqrt{-\frac{g_{44}}{g_{22}}} = 1 - \frac{\alpha}{2r}\left(1 + \frac{x_2^2}{r^2}\right).$$

Carrying out the calculation, this gives

$$B = \frac{2\alpha}{\Delta} = \frac{\kappa M}{2\pi\Delta}. \tag{74}$$

According to this, a ray of light going past the sun undergoes a deflection of 1.7″; and a ray going past the planet Jupiter a deflection of about 0.02″.

If we calculate the gravitational field to a higher degree of approximation, and likewise with corresponding accuracy the orbital motion of a material point of relatively infinitely small mass, we find a deviation of the following kind from the Kepler-Newton laws of planetary motion. The orbital ellipse of a planet undergoes a slow rotation, in the direction of motion, of amount

$$\epsilon = 24\pi^3 \frac{a^2}{T^2 c^2 (1 - e^2)} \tag{75}$$

per revolution. In this formula a denotes the major semi-axis, c the velocity of light in the usual measurement, e the eccentricity, T the time of revolution in seconds.[19]

Calculation gives for the planet Mercury a rotation of the orbit of 43″ per century, corresponding exactly to astronomical observation (Leverrier); for the astronomers have discovered in the motion of the perihelion of this planet, after allowing for disturbances by other planets, an inexplicable remainder of this magnitude.

---

[19] For the calculation I refer to the original papers: A. Einstein, Sitzungsber. d. Preuss. Akad. d. Wiss., 1915, p. 831; K. Schwarzschild, ibid., 1916, p. 189.

Part III

BIRTH OF A DYNAMICAL UNIVERSE

Cosmological Considerations on the General Theory of Relativity

A. Einstein

Translation of Kosmologische Betrachtungen zur allgemeinen Relativitätstheorie, *Sitzungsberichte der Preussischen Akad. d. Wissenschaften*, 1917. Original English publication in: H. A. Lorentz, A. Einstein, H. Minkowski and H. Weyl, *The Principle of Relativity: A Collection of Original Memoirs on the Special and General Theory of Relativity*. With Notes by A. Sommerfeld. Translated by W. Perrett and G. B. Jeffery (Methuen and Company, Ltd., 1923; reprinted by Dover Publications Inc., 1952).

It is well known that Poisson's equation

$$\nabla^2 \phi = 4\pi K \rho \tag{1}$$

in combination with the equations of motion of a material point is not as yet a perfect substitute for Newton's theory of action at a distance. There is still to be taken into account the condition that at spatial infinity the potential ϕ tends toward a fixed limiting value. There is an analogous state of things in the theory of gravitation in general relativity. Here, too, we must supplement the differential equations by limiting conditions at spatial infinity, if we really have to regard the universe as being of infinite spatial extent.

In my treatment of the planetary problem I chose these limiting conditions in the form of the following assumption: it is possible to select a system of reference so that at spatial infinity all the gravitational potentials $g_{\mu\nu}$ become constant. But it is by no means evident *a priori* that we may lay down the same limiting conditions when we wish to take larger portions of the physical universe into consideration. In the following pages the reflections will be given which, up to the present, I have made on this fundamentally important question.

§ 1. The Newtonian Theory

It is well known that Newton's limiting condition of the constant limit for ϕ at spatial infinity leads to the view that the density of matter becomes zero at infinity. For we imagine that there may be a place in universal space round about which the gravitational field of matter, viewed on a large scale, possesses spherical symmetry.

181

It then follows from Poisson's equation that, in order that ϕ may tend to a limit at infinity, the mean density ρ must decrease toward zero more rapidly than $1/r^2$ as the distance r from the centre increases.[1] In this sense, therefore, the universe according to Newton is finite, although it may possess an infinitely great total mass.

From this it follows in the first place that the radiation emitted by the heavenly bodies will, in part, leave the Newtonian system of the universe, passing radially outwards, to become ineffective and lost in the infinite. May not entire heavenly bodies fare likewise? It is hardly possible to give a negative answer to this question. For it follows from the assumption of a finite limit for ϕ at spatial infinity that a heavenly body with finite kinetic energy is able to reach spatial infinity by overcoming the Newtonian forces of attraction. By statistical mechanics this case must occur from time to time, as long as the total energy of the stellar system – transferred to one single star – is great enough to send that star on its journey to infinity, whence it never can return.

We might try to avoid this peculiar difficulty by assuming a very high value for the limiting potential at infinity. That would be a possible way, if the value of the gravitational potential were not itself necessarily conditioned by the heavenly bodies. The truth is that we are compelled to regard the occurrence of any great differences of potential of the gravitational field as contradicting the facts. These differences must really be of so low an order of magnitude that the stellar velocities generated by them do not exceed the velocities actually observed.

If we apply Boltzmann's law of distribution for gas molecules to the stars, by comparing the stellar system with a gas in thermal equilibrium, we find that the Newtonian stellar system cannot exist at all. For there is a finite ratio of densities corresponding to the finite difference of potential between the centre and spatial infinity. A vanishing of the density at infinity thus implies a vanishing of the density at the centre.

It seems hardly possible to surmount these difficulties on the basis of the Newtonian theory. We may ask ourselves the question whether they can be removed by a modification of the Newtonian theory. First of all we will indicate a method which does not in itself claim to be taken seriously; it merely serves as a foil for what is to follow. In place of Poisson's equation we write

$$\nabla^2 \phi - \lambda\phi = 4\pi\kappa\rho, \tag{2}$$

where λ denotes a universal constant. If ρ_0 be the uniform density of a distribution of mass, then

$$\phi = -\frac{4\pi\kappa}{\lambda}\rho_0 \tag{3}$$

is a solution of equation (2). This solution would correspond to the case in which the matter of the fixed stars was distributed uniformly through space, if the density ρ_0 is equal to the actual mean density of the matter in the universe. The solution then corresponds to an infinite extension of the central space, filled uniformly with matter.

---

[1] ρ is the mean density of matter, calculated for a region which is large as compared with the distance between neighbouring fixed stars, but small in comparison with the dimensions of the whole stellar system.

If, without making any change in the mean density, we imagine matter to be non-uniformly distributed locally, there will be, over and above the ϕ with the constant value of equation (3), an additional ϕ, which in the neighbourhood of denser masses will so much the more resemble the Newtonian field as $\lambda\phi$ is smaller in comparison with $4\pi\kappa\rho$.

A universe so constituted would have, with respect to its gravitational field, no centre. A decrease of density in spatial infinity would not have to be assumed, but both the mean potential and mean density would remain constant to infinity. The conflict with statistical mechanics which we found in the case of the Newtonian theory is not repeated. With a definite but extremely small density, matter is in equilibrium, without any internal material forces (pressures) being required to maintain equilibrium.

§ 2. The Boundary Conditions According to the General Theory of Relativity

In the present paragraph I shall conduct the reader over the road that I have myself travelled, rather a rough and winding road, because otherwise I cannot hope that he will take much interest in the result at the end of the journey. The conclusion I shall arrive at is that the field equations of gravitation which I have championed hitherto still need a slight modification, so that on the basis of the general theory of relativity those fundamental difficulties may be avoided which have been set forth in § 1 as confronting the Newtonian theory. This modification corresponds perfectly to the transition from Poisson's equation (1) to equation (2) of § 1. We finally infer that boundary conditions in spatial infinity fall away altogether, because the universal continuum in respect of its spatial dimensions is to be viewed as a self-contained continuum of finite spatial (three-dimensional) volume.

The opinion which I entertained until recently, as to the limiting conditions to be laid down in spatial infinity, took its stand on the following considerations. In a consistent theory of relativity there can be no inertia *relatively* to "*space*" but only an inertia of masses *relatively to one another*.[2] If, therefore, I have a mass at a sufficient

---

[2]EDITOR'S NOTE: Here, clearly, Einstein is still under the influence of Mach's ideas and still not fully appreciating Minkowski's results. As inertia is the resistance a mass (a body) offers to its acceleration, what Einstein is saying is that there can be no acceleration *relatively* to "*space*" but only an acceleration of masses *relatively to one another*. Einstein's belief that acceleration is relative is at odds with Minkowski's explanation that acceleration is represented by an absolute geometrical property of the worldline of an accelerating particle – its curvature (i.e., deformation) – and is therefore absolute in spacetime physics; see EDITOR'S NOTES on pp. 128 and 138. One can better understand Minkowski's remark – "Especially the concept of *acceleration* acquires a sharply prominent character" (this volume, p. 112) – by noticing that that "prominent character" of acceleration reveals why acceleration is absolute in spacetime: not because it is with respect to an absolute space, but because it is a manifestation (in terms of our everyday three-dimensional language) of the curvature (deformation) of the worldtube of an accelerating particle, which is an absolute geometrical property, reflecting the experimental fact that an accelerating particle *resists* its acceleration (in full agreement with the assumption that the deformed worldtube of an accelerating particle should resist its deformation; again, unlike Poincaré, Minkowski regarded spacetime and the worldtubes of particles as representing the real physical world and therefore a real worldtube is indeed expected to resist its deformation). Independently, it becomes crystal clear why acceleration

184

distance from all other masses in the universe, its inertia must fall to zero.³ We will try to formulate this condition mathematically.

According to the general theory of relativity the negative momentum is given by the first three components, the energy by the last component of the covariant tensor multiplied by $\sqrt{-g}$

$$m\sqrt{-g}\, g_{\mu\nu}\frac{dx_\alpha}{ds} \tag{4}$$

where, as always, we set

$$ds^2 = g_{\mu\nu}dx_\mu dx_\nu. \tag{5}$$

In the particularly perspicuous case of the possibility of choosing the system of coordinates so that the gravitational field at every point is spatially isotropic, we have more simply

$$ds^2 = -A(dx_1^2 + dx_2^2 + dx_3^2) + Bdx_4^2.$$

If, moreover, at the same time

$$\sqrt{-g} = 1 = \sqrt{A^3 B},$$

we obtain from (4), to a first approximation for small velocities

$$m\frac{A}{\sqrt{B}}\frac{dx_1}{dx_4}, \qquad m\frac{A}{\sqrt{B}}\frac{dx_2}{dx_4}, \qquad m\frac{A}{\sqrt{B}}\frac{dx_3}{dx_4}$$

for the components of momentum, and for the energy (in the static case)

$$m\sqrt{B}.$$

From the expressions for the momentum, it follows that $m\frac{A}{\sqrt{B}}$ plays the part of the rest mass. As m is a constant peculiar to the point of mass, independently of its position, this expression, if we retain the condition $\sqrt{-g} = 1$ at spatial infinity, can vanish only when A diminishes to zero, while B increases to infinity. Such a degeneration of the coefficients $g_{\mu\nu}$ seems to be required by the postulate of the relativity of all inertia.⁴ This requirement implies that the potential energy $m\sqrt{B}$ becomes infinitely great at infinity. Thus a point of mass can never leave the system; and a more detailed investigation shows that the same thing applies to light-rays. A system of the universe with such behaviour of the gravitational potentials at infinity would not therefore run the risk of wasting away which was mooted just now in connection with the Newtonian theory.

is absolute not because it is with respect to an absolute space – Minkowski showed that there is no such thing as absolute space in the physical world; what exists is an absolute four-dimensional world, which can be *described* by observers in relative motion in terms of their spaces and times.

³EDITOR'S NOTE: As indicated in the above footnote, it is the absolute geometrical property of the worldtube of the mass – its curvature, i.e., its deformation – that represents the absolute acceleration of the mass and, therefore, its inertia. That is why, if that *single* mass (far away of other masses) is absolutely accelerating, which means that its worldtube is deformed, its inertia will be exactly equal to the inertia it exhibits in flat spacetime (assuming that "far away of other masses" means the mass in question is in flat spacetime).

⁴EDITOR'S NOTE: Here, Einstein's analysis led him to use the expression "seems to be required" carefully.

I wish to point out that the simplifying assumptions as to the gravitational potentials on which this reasoning is based, have been introduced merely for the sake of lucidity. It is possible to find general formulations for the behaviour of the $g_{\mu\nu}$ at infinity which express the essentials of the question without further restrictive assumptions.

At this stage, with the kind assistance of the mathematician J. Grommer, I investigated centrally symmetrical, static gravitational fields, degenerating at infinity in the way mentioned. The gravitational potentials $g_{\mu\nu}$ were applied, and from them the energy-tensor $T_{\mu\nu}$ of matter was calculated on the basis of the field equations of gravitation. But here it proved that for the system of the fixed stars no boundary conditions of the kind can come into question at all, as was also rightly emphasized by the astronomer de Sitter recently.

For the contravariant energy-tensor $T^{\mu\nu}$ of ponderable matter is given by

$$T^{\mu\nu} = \rho \frac{dx_\mu}{ds} \frac{dx_\nu}{ds},$$

where ρ is the density of matter in natural measure. With an appropriate choice of the system of coordinates the stellar velocities are very small in comparison with that of light. We may, therefore, substitute $\sqrt{g_{44}}\, dx_4$ for ds. This shows us that all components of $T^{\mu\nu}$ must be very small in comparison with the last component T^{44}. But it was quite impossible to reconcile this condition with the chosen boundary conditions. In the retrospect this result does not appear astonishing. The fact of the small velocities of the stars allows the conclusion that wherever there are fixed stars, the gravitational potential (in our case $\sqrt{B}$) can never be much greater than here on earth. This follows from statistical reasoning, exactly as in the case of the Newtonian theory. At any rate, our calculations have convinced me that such conditions of degeneration for the $g_{\mu\nu}$ at spatial infinity may not be postulated.

After the failure of this attempt, two possibilities next present themselves.

(a) We may require, as in the problem of the planets, that, with a suitable choice of the system of reference, the $g_{\mu\nu}$ at spatial infinity approximate to the values

$$
\begin{array}{cccc}
1 & 0 & 0 & 0 \\
0 & 1 & 0 & 0 \\
0 & 0 & 1 & 0 \\
0 & 0 & 0 & -1
\end{array}
$$

(b) We may refrain entirely from laying down boundary conditions for spatial infinity claiming general validity; but at the spatial limit of the domain under consideration we have to give the $g_{\mu\nu}$ separately in each individual case, as hitherto we were accustomed to give the initial conditions for time separately.

The possibility (b) holds out no hope of solving the problem, but amounts to giving it up. This is an incontestable position, which is taken up at the present time by de Sitter.[5] But I must confess that such a complete resignation in this fundamental question is for me a difficult thing. I should not make up my mind to it until every effort to make headway toward a satisfactory view had proved to be vain.

---

[5] de Sitter, Akad. van Wetensch. te Amsterdam, 8 Nov., 1916.

Possibility (a) is unsatisfactory in more respects than one. In the first place those boundary conditions presuppose a definite choice of the system of reference, which is contrary to the spirit of the relativity principle. Secondly, if we adopt this view, we fail to comply with the requirement of the relativity of inertia. For the inertia of a material point of mass m (in natural measure) depends upon the $g_{\mu\nu}$; but these differ but little from their postulated values, as given above, for spatial infinity. Thus inertia would indeed be *influenced* , but would not be *conditioned* by matter (present in finite space). If only one single point of mass were present, according to this view, it would possess inertia, and in fact an inertia almost as great as when it is surrounded by the other masses of the actual universe. Finally, those statistical objections must be raised against this view which were mentioned in respect of the Newtonian theory.

From what has now been said it will be seen that I have not succeeded in formulating boundary conditions for spatial infinity. Nevertheless, there is still a possible way out, without resigning as suggested under (b). For if it were possible to regard the universe as a continuum which is *finite (closed) with respect to its spatial dimensions*, we should have no need at all of any such boundary conditions. We shall proceed to show that both the general postulate of relativity and the fact of the small stellar velocities are compatible with the hypothesis of a spatially finite universe; though certainly, in order to carry through this idea, we need a generalizing modification of the field equations of gravitation.

§ 3. The Spatially Finite Universe with a Uniform Distribution of Matter

According to the general theory of relativity the metrical character (curvature) of the four-dimensional space-time continuum is defined at every point by the matter at that point and the state of that matter. Therefore, on account of the lack of uniformity in the distribution of matter, the metrical structure of this continuum must necessarily be extremely complicated. But if we are concerned with the structure only on a large scale, we may represent matter to ourselves as being uniformly distributed over enormous spaces, so that its density of distribution is a variable function which varies extremely slowly. Thus our procedure will somewhat resemble that of the geodesists who, by means of an ellipsoid, approximate to the shape of the earth's surface, which on a small scale is extremely complicated.

The most important fact that we draw from experience as to the distribution of matter is that the relative velocities of the stars are very small as compared with the velocity of light. So I think that for the present we may base our reasoning upon the following approximative assumption. There is a system of reference relatively to which matter may be looked upon as being permanently at rest. With respect to this system, therefore, the contravariant energy-tensor $T^{\mu\nu}$ of matter is, by reason of (5), of the simple form

$$\begin{matrix} 0 & 0 & 0 & 0 \\ 0 & 0 & 0 & 0 \\ 0 & 0 & 0 & 0 \\ 0 & 0 & 0 & \rho \end{matrix} \tag{6}$$

The scalar ρ of the (mean) density of distribution may be *a priori* a function of the space coordinates. But if we assume the universe to be spatially finite, we are prompted to the hypothesis that ρ is to be independent of locality. On this hypothesis we base the following considerations.

As concerns the gravitational field, it follows from the equation of motion of the material point

$$\frac{d^2 x_\nu}{ds^2} + \{\alpha\beta, \nu\} \frac{dx_\alpha}{ds} \frac{dx_\beta}{ds} = 0$$

that a material point in a static gravitational field can remain at rest only when g_{44} is independent of locality. Since, further, we presuppose independence of the time coordinate x_4 for all magnitudes, we may demand for the required solution that, for all x_ν,

$$g_{44} = 1. \tag{7}$$

Further, as always with static problems, we shall have to set

$$g_{14} = g_{24} = g_{34} = 0. \tag{8}$$

It remains now to determine those components of the gravitational potential which define the purely spatial-geometrical relations of our continuum $(g_{11}, g_{12}, \ldots g_{33})$. From our assumption as to the uniformity of distribution of the masses generating the field, it follows that the curvature of the required space must be constant. With this distribution of mass, therefore, the required finite continuum of the x_1, x_2, x_3, with constant x_4 will be a spherical space.

We arrive at such a space, for example, in the following way. We start from a Euclidean space of four dimensions, $\xi_1, \xi_2, \xi_3, \xi_4$ with a linear element $d\sigma$; let, therefore,

$$d\sigma^2 = d\xi_1^2 + d\xi_2^2 + d\xi_3^2 + d\xi_4^2. \tag{9}$$

In this space we consider the hypersurface

$$R^2 = \xi_1^2 + \xi_2^2 + \xi_3^2 + \xi_4^2, \tag{10}$$

where R denotes a constant. The points of this hypersurface form a three-dimensional continuum, a spherical space of radius of curvature R.

The four-dimensional Euclidean space with which we started serves only for a convenient definition of our hypersurface. Only those points of the hypersurface are of interest to us which have metrical properties in agreement with those of physical space with a uniform distribution of matter. For the description of this three-dimensional continuum we may employ the coordinates ξ_1, ξ_2, ξ_3 (the projection upon the hyperplane $\xi_4 = 0$) since, by reason of (10), ξ_4 can be expressed in terms of ξ_1, ξ_2, ξ_3. Eliminating ξ_4 from (9), we obtain for the linear element of the spherical space the expression

$$d\sigma^2 = \gamma_{\mu\nu} d\xi_\mu d\xi_\nu$$
$$\gamma_{\mu\nu} = \delta_{\mu\nu} + \frac{\xi_\mu \xi_\nu}{R^2 - \rho^2} \tag{11}$$

188

where $\delta_{\mu\nu} = 1$, if $\mu = \nu$; $\delta_{\mu\nu} = 0$, if $\mu \neq \nu$, and $\rho^2 = \xi_1^2 + \xi_2^2 + \xi_3^2$. The coordinates chosen are convenient when it is a question of examining the environment of one of the two points $\xi_1 = \xi_2 = \xi_3 = 0$.

Now the linear element of the required four-dimensional space-time universe is also given us. For the potential $g_{\mu\nu}$, both indices of which differ from 4, we have to set

$$g_{\mu\nu} = -\left(\delta_{\mu\nu} + \frac{x_\mu x_\nu}{R^2 - (x_1^2 + x_2^2 + x_3^2)}\right), \tag{12}$$

which equation, in combination with (7) and (8), perfectly defines the behaviour of measuring-rods, clocks, and light-rays.

§ 4. On an Additional Term for the Field Equations of Gravitation

My proposed field equations of gravitation for any chosen system of coordinates run as follows:

$$G_{\mu\nu} = -\kappa\left(T_{\mu\nu} - \frac{1}{2}g_{\mu\nu}T\right),$$

$$G_{\mu\nu} = -\frac{\partial}{\partial x_\alpha}\{\mu\nu, \alpha\} + \{\mu\alpha, \beta\}\{\nu\beta, \alpha\} \tag{13}$$

$$+ \frac{\partial^2 \log\sqrt{-g}}{\partial x_\mu \partial x_\nu} - \{\mu\nu, \alpha\}\frac{\partial \log\sqrt{-g}}{\partial x_\alpha}.$$

The system of equations (13) is by no means satisfied when we insert for the $g_{\mu\nu}$ the values given in (7), (8), and (12), and for the (contravariant) energy-tensor of matter the values indicated in (6). It will be shown in the next paragraph how this calculation may conveniently be made. So that, if it were certain that the field equations (13) which I have hitherto employed were the only ones compatible with the postulate of general relativity, we should probably have to conclude that the theory of relativity does not admit the hypothesis of a spatially finite universe.

However, the system of equations (13) allows a readily suggested extension which is compatible with the relativity postulate, and is perfectly analogous to the extension of Poisson's equation given by equation (2). For on the left-hand side of field equation (13) we may add the fundamental tensor $g_{\mu\nu}$ multiplied by a universal constant, $-\lambda$, at present unknown, without destroying the general covariance. In place of field equation (13) we write

$$G_{\mu\nu} - \lambda g_{\mu\nu} = -\kappa(T_{\mu\nu} - \frac{1}{2}g_{\mu\nu}T). \tag{13a}$$

This field equation, with λ sufficiently small, is in any case also compatible with the facts of experience derived from the solar system. It also satisfies laws of conservation of momentum and energy, because we arrive at (13a) in place of (13) by introducing into Hamilton's principle, instead of the scalar of Riemann's tensor, this scalar increased by a universal constant; and Hamilton's principle, of course, guarantees the validity of laws of conservation. It will be shown in § 5 that field equation (13a) is compatible with our conjectures on field and matter.

§ 5. Calculation and Result

Since all points of our continuum are on an equal footing, it is sufficient to carry through the calculation for one point, e.g. for one of the two points with the coordinates

$$x_1 = x_2 = x_3 = x_4 = 0.$$

Then for the $g_{\mu\nu}$ in (13a) we have to insert the values

$$
\begin{array}{cccc}
-1 & 0 & 0 & 0 \\
0 & -1 & 0 & 0 \\
0 & 0 & -1 & 0 \\
0 & 0 & 0 & 1
\end{array}
$$

We thus obtain in the first place

$$G_{\mu\nu} = \frac{\partial}{\partial x_1}\,[\mu\nu,\,1] + +\frac{\partial}{\partial x_2}\,[\mu\nu,\,2] + \frac{\partial}{\partial x_3}\,[\mu\nu,\,3] + \frac{\partial^2 \log \sqrt{-g}}{\partial x_\mu \partial x_\nu}.$$

From this we readily discover, taking (7), (8), and (13) into account, that all equations (13a) are satisfied if the two relations

$$-\frac{2}{R^2} + \lambda = -\frac{\kappa\rho}{2}, \qquad -\lambda = -\frac{\kappa\rho}{2},$$

or

$$\lambda = \frac{\kappa\rho}{2} = \frac{1}{R^2} \tag{14}$$

are fulfilled.

Thus the newly introduced universal constant λ defines both the mean density of distribution ρ which can remain in equilibrium and also the radius R and the volume $2\pi^2 R^3$ of spherical space. The total mass M of the universe, according to our view, is finite, and is in fact

$$M = \rho\,2\pi^2 R^3 = 4\pi^2 \frac{R}{\kappa} = \pi\sqrt{\frac{32}{\kappa^3\rho}}. \tag{15}$$

Thus the theoretical view of the actual universe, if it is in correspondence with our reasoning, is the following. The curvature of space is variable in time and place, according to the distribution of matter, but we may roughly approximate to it by means of a spherical space. At any rate, this view is logically consistent, and from the standpoint of the general theory of relativity lies nearest at hand; whether, from the standpoint of present astronomical knowledge, it is tenable, will not here be discussed. In order to arrive at this consistent view, we admittedly had to introduce an extension of the field equations of gravitation which is not justified by our actual knowledge of gravitation. It is to be emphasized, however, that a positive curvature of space is given by our results, even if the supplementary term is not introduced. That term is necessary only for the purpose of making possible a quasi-static distribution of matter, as required by the fact of the small velocities of the stars.

ON THE CURVATURE OF SPACE

W. DE SITTER

W. de Sitter, On the curvature of space, in: *KNAW, Proceedings*, 20 I, 1918, Amsterdam, 1918, pp. 229-243 (Communicated in the meeting of 1917, June 30)

1. In order to make possible an entirely relative conception of inertia, Einstein[1] has replaced the orlginal field equations of his theory by the equations

$$G_{\mu\nu} - \frac{1}{2}g_{\mu\nu}\lambda = -\kappa T_{\mu\nu} + \frac{1}{2}\kappa g_{\mu\nu}T. \tag{1}$$

In my last paper[2] I have pointed out two different systems of $g_{\mu\nu}$ which satisfy these equations. The system A is Einstein's, in which the whole of space is filled with matter of the average density ρ_0. In a stationary state, and If all matter is at rest without any stresses or pressure, then we have $T_{\mu\nu} = 0$, with the exception of $T_{44} = g_{44}\rho_0$. In the system B this "world-matter" does not exist: we have $\rho_0 = 0$ and consequently all $T_{\mu\nu} = 0$. The line element in the two systems was there found to be

$$ds^2 = -R^2 \left\{ d\chi^2 + \sin^2\chi \left[d\psi^2 + \sin^2\psi\, d\theta^2 \right] \right\} + c^2 dt^2, \tag{2A}$$

$$ds^2 = -R^2 \left\{ d\omega^2 + \sin^2\omega \left[d\chi^2 + \sin^2\chi(d\psi^2 + \sin^2\psi\, d\theta^2) \right] \right\}. \tag{2B}$$

In the system A we have:

$$\lambda = \frac{1}{R^2}, \qquad \kappa\rho_0 = 2\lambda, \tag{3A}$$

and in B:

$$\lambda = \frac{3}{R^2}, \qquad \rho_0 = 0. \tag{3B}$$

---

[1] A. Einstein, *Kosmologishe Betrachtungen zur Allgemeinen Relativitätstheorie*, Sitzungsber., Berlin 1917 Febr. 8 p. 142.

[2] W. de Sitter, *On the relativity of inertia*, these Proceedings, 1917 March 31, vol. XIX p. 1217. (EDITOR'S NOTE: By "these Proceedings" de Sitter means KNAW, Proceedings, where this paper (On the Curvature of Space) is published.)
In the footnote to page 1220 of that paper it is stated that the four-dimensional world of the system B can be represented as a hyperboloid of two sheets in a space of five dimensions, which is projected on a Euclidean space of four-dimensions by a "stereographic projection". This is erroneous. The hyperboloid has only one sheet. Its projection fills only part of the Euclidean space of four dimensions; the part outside the limiting hyperboloid $1 + \sigma h^2 = 0$ (which is called (a) in the quoted footnote) is the projection of the conjugated hyperboloid (which is of two sheets).

In the system A χ, ψ, θ are real angles; in B ψ and θ are also real, but ω and χ are imaginary. If, however, we put

$$\sin \omega \sin \chi = \sin \zeta, \qquad r = R\zeta$$
$$\tan \omega \cos \chi = \tan i\eta, \qquad t = R\eta,$$

where $i = \sqrt{-1}$, then ξ and η are real and (2B) becomes:

$$ds^2 = -dr^2 - R^2 \sin^2 \frac{r}{R} \left[d\psi^2 + \sin^2 \psi \, d\theta^2 \right] + \cos^2 \frac{r}{R} c^2 dt^2. \qquad (4B)$$

If in A we also take $r = R\chi$, then (2A) becomes

$$ds^2 = -dr^2 - R^2 \sin^2 \frac{r}{R} \left[d\psi^2 + \sin^2 \psi \, d\theta^2 \right] + c^2 dt^2. \qquad (4A)$$

The two systems A and B now differ only in g_{44}. For the sake of comparison we add the system C, with

$$\lambda = 0, \qquad \rho_0 = 0, \qquad (3C)$$

in which the line-element is

$$ds^2 = -dr^2 - r^2 [d\psi^2 + \sin^2 \psi \, d\theta^2] + c^2 dt^2. \qquad (4C)$$

Both A and B become identical with C for $R = \infty$.

If in A the origin of coordinates is displaced to a point χ_1, ψ_1, θ_1 and in B to a time-space point $\omega_1, \chi_1, \psi_1, \theta_1$, then the line-element conserves the forms (2A) and (2B) respectively. These can then again by the same transformations be altered to (4A) and (4B). In A the variable t, which takes no part in the transformation, remains of course the same. In B on the other hand the new variable t after the transformation is generally not the same as before.

I will put, for both systems A and B

$$\chi = \frac{r}{R}.$$

In the system B this χ is not the same as in (2B), but it is the angle which was called ζ above, I will continue to use r as an independent variabie, and not χ.

2. In the theory of general relativity there is no essential difference between inertia and gravitation. It will, however, be convenient to continue to make this difference. A field in which the line-element can be brought in one of the forms (4A), (4B) or (4C) with the corresponding condition (3A), (3B), or (3C), will be called a field of pure inertia, without gravitation. If the $g_{\mu\nu}$ deviate from these values we will say that there is gravitation, This is produced by matter, which I call "ordinary" or "gravitating" matter. Its density is ρ_1. In the systems B and C there is no other matter than this ordinary matter. In the system A the whole of space is filled with matter, which, in the simple case that the line-element is represented by (2A) or (4 A) produces no "gravitation", but only "inertia". This matter I have called "world-matter". Its density is ρ_0. When taken over sufficiently large units of volume this ρ_0

is a constant. Locally however it may be variable the world-matter can be condensed to bodies of greater density, or it can have a smaller density than the average, or be absent altogether. According to Einstein's view we must assume that *all* ordinary matter (sun, stars, nebulae etc.) consist of condensed world-matter, and perhaps also that all world-matter is thus condensed.

3. To begin with we will neglect gravitation and consider only the inertial field. The three-dimensional line-element is in the two systems A and B.

$$d\sigma^2 = dr^2 + R^2 \sin^2 \frac{r}{R} \left[d\psi^2 + \sin^2 \psi \, d\theta^2 \right].$$

If R^2 is positive and finite, this is the line-element of a three-dimensional space with a constant positive curvature. There are two forms of this, viz: the space of Riemann,[3] or *spherical space*, and the elliptical space, which has been investigated by Newcomb.[4] In the spherical space all "straight" (i.e. geodesic) lines which start from one point, intersect again in another point: the "antipodal point", whose distance from the fist point, measured along any of these lines, is πR. In the elliptical space any two straight lines have only one point in common. In both spaces the straight line is closed; in the spherical space its total length is $2\pi R$, in the elliptical space it is πR. In the spherical space the largest possible distance between two points is πR, in the elliptical space $\frac{1}{2}\pi R$. Both spaces are finite, though unlimited. The volume of the whole of spherical space is $2\pi^2 R^3$, of elliptical space $\pi^2 R^3$. For values of r which are small compared with R, the two spaces differ only inappreciably form the Euclidean space.

The existence of the antipodal point, where all rays of light starting from a point again intersect, and where also, as will be shown below, the gravitational action of a material point (however small its mass may be) becomes infinite, certainly is a drawback of the spherical space, and it will be preferable to assume the true physical space to be elliptical.

The elliptical space can be projected on Euclidean space by the transformation

$$r = R \tan \chi. \tag{5}$$

The line-element in the systems A and B then becomes

$$ds^2 = -\frac{dr^2}{\left(1 + \frac{r^2}{R^2}\right)^2} - \frac{r^2[d\psi^2 + \sin^2 \psi \, d\theta^2]}{1 + \frac{r^2}{R^2}} + c^2 dt^2. \tag{6A}$$

$$ds^2 = -\frac{dr^2}{\left(+ \frac{r^2}{R^2}\right)^2} - \frac{r^2[d\psi^2 + \sin^2 \psi \, d\theta^2]}{1 + \frac{r^2}{R^2}} + \frac{c^2 dt^2}{1 + \frac{r^2}{R^2}}. \tag{6B}$$

For $r = \infty$ in the system A all $g_{\mu\nu}$ become zero, with the exception of g_{44} which remains 1. In the system B g_{44} also becomes zero.

---

[3] *Ueber die Hypothesen welche der Geometrie zu Grunde liegen* (1854).

[4] *Elementary theorems relating to geometry of three dimensions and of uniform positive curvature*, Crelle's Journal Bd, 83, p. 293 (1877).

4. The world-lines of light-vibrations are geodesic lines ($ds = 0$) in the four-dimensional time-space. Their projections on the the three-dimensional space are the rays of light. In the system A, with the coordinates r, ψ, θ, these light-rays are also geodesic lines of the three-dimensional space, and the velocity of light is constant. In the system B this is not so. The velocity of light in that system is, in the radial direction, $v = c \cos \chi$. It is possible, however, in B to introduce space-coordinates, measured in which the velocity of light shall be constant in the radial direction. If the radius-vector in this new measure is called h, we have

$$\cos \chi \, dh = dr.$$

The integral of this equation is

$$\sinh \frac{h}{R} = \tan \frac{r}{R}. \tag{7}$$

In the system A we can, of course, also perform the same transformation. The line-element becomes

$$ds^2 = \frac{-dh^2 - \sinh^2 \frac{h}{R} \left[d\psi^2 + \sin^2 \psi \, d\theta^2 \right]}{\cosh^2 \frac{h}{R}} + c^2 dt^2 \tag{8A}$$

$$ds^2 = \frac{-dh^2 - \sinh^2 \frac{h}{R} \left[d\psi^2 + \sin^2 \psi \, d\theta^2 \right] + c^2 dt^2}{\cosh^2 \frac{h}{R}}. \tag{8B}$$

The three-dimensional line-element

$$d\sigma^2 = dh^2 \sinh^2 \frac{h}{R} \left[d\psi^2 + \sin^2 \psi \, d\theta^2 \right]$$

is that of a space of constant negative curvature: the *hyperbolical space* or space of Lobachevsky. When described in the coordinates of this space, the rays of light in the system B are straight (i.e. geodesic) lines, and the velocity of light is constant in all directions, although the system of reference was determined by the condition that it should be constant in the radial direction.

In this system of reference also all $g_{\mu\nu}$ are zero at infinity in the system B, and in A all $g_{\mu\nu}$ excepting g_{44}, which remains 1.

To $h = \infty$ corresponds $r = \frac{1}{2}\pi R$. The whole of eliptical space is therefore by the transformation (7) projected on the whole of hyperbolical space. For values of r exceeding $\frac{1}{2}\pi R$, h becomes negative. Now a point $(-h, \psi, \theta)$, is the same as $(h, \pi - \psi, \pi + \theta)$. The projection of the spherical space therefore fills the hyperbolical space twice. Tbe same thing is true of the projection, by (5), of the elliptical and spherical spaces on the Euclidean space.

5. Let the sun be placed at the origin of coordinates, and let the distance from the sun to the earth be a. We still neglect all gravitation.

In the system A the rays of light are straight lines, when described in the coordinates r, ψ, θ, i.e. in the elliptical or spherical space.

In the system B the same is true for the coordinates h, ψ, θ (hyperbolical space).

In the system A, consequently to triangles formed by rays of light, the ordinary formulas of spherical trigonometry are applicable. The parallax p of a star whose distance from the sun is r, is thus given by the formula

$$\tan p = \sin \frac{a}{R} \cot \frac{r}{R}.$$

The square of a/R being negligible, we can write this

$$p = \frac{a}{R} \cot \frac{r}{R} = \frac{a}{r}. \tag{9A}$$

In the system B we have similarly, in the coordinates h, ψ, θ:

$$\tan p = \sinh \frac{a}{R} \coth \frac{h}{R},$$

or

$$p = \frac{a}{R} \coth \frac{h}{R} = \frac{a}{R \sin \chi} = \frac{a}{r} \sqrt{1 + \frac{r^2}{R^2}}. \tag{9B}$$

In the system A we have consequently $p = 0$ for $r = \frac{1}{2}\pi R$ i.e. for the largest distance which is possible in the elliptical space. If we admitted still larger distances, which are only possible in the spherical space, then p would become negative, and for $r = \pi R$ we should find $p = -90°$.

In the system B p has a minimum value

$$p_0 = \frac{a}{R},$$

which it reaches for $h = \infty$, i.e. $r = \frac{1}{2}\pi R$. For values of r exceeding this distance p increases again, and for $r = \pi R$ we should find $p = +90°$.

Already in 1900 Schwardschild[5] gave a discussion of the possible curvature of space, starting from the formulae (9A) and (9B). For the system B we can from the observed parallaxes[6] derive a lower limit for R. Schwardschild finds $R > 4.10^6$ astronomical units. In the system A the measured parallaxes cannot give a limit for R.

In both systems we can, of course, derive such a limit from *distances* which have been determined, or estimated, otherwise than from the measured parallaxes. These distances must in the elliptical space, be smaller than $\frac{1}{2}\pi R$. This undoubtedly leads to a much higher limit, of the order of 10^{10} or more.

6.The straight line being closed, we should, at the point of the heavens 180° from the sun, see an image of the back side of the sun. This not being the case,

---

[5] *Ueber das zulassige Krimmungsmaass des Raumes*, Vierteljahrsschrift der Astron Gesellschaft, Bd. 35 p. 337.

[6] The meaning is of course actually measured parallaxes, not parallaxes derived – by the formula $p = a/r$ from a distance which is determined from other sources (comparison of radial and transversal velocity, absolute magnitude, etc.). Schwardschild assumes that there are certainly stars having a parallax of $0''05$. All parallaxes measured since then are *relative* parallaxes, and consequently we must at the present time still use the same limit.

practically all the light must be absorbed on the long "voyage round the universe". Schwardschild estimates that an absorption of 40 magnitudes would be sufficient.[7] If we adopt the result found by Shapley,[8] viz. that the absorption in the intergalactic space is smaller than $0^m.01$ in a distance of 1000 parsecs, then for an absorption of 40 mags we need a distance of 7.10^{11} astronomical units. In the elliptical space we have thus $R > \frac{1}{4} \times 10^{12}$.

In the system A we can suppose that this absorption is produced by the world-matter. It is about 1/50 of the absorption which King[9] used in his calculation of the density of matter in interstellar space. The density of the world-matter would thus be about 1/50 of the density found by King, or $\rho_0 = \frac{3}{2} \times 10^{-14}$ in astronomical units. The corresponding value of R (see art. 8) is $R = 2 \times 10^{10}$. The total absorption in the distance πR would then be only 3.6 magnitudes. To get the required absorption of 40 magnitudes we must increase ρ_0 and consequently diminish R. We then find $\rho_0 = 2 \times 10^{-12}$, $R = 2 \times 10^9$ This value of course has practically no weight, as it is very doubtful whether the considerations by which King derived the density from the coefficient of absorption are applicable to the world-matter.

The whole argument is inapplicable to the system B, since in this system the light requires an infinite time for the "voyage round the world." One half of this time is

$$T = \int_0^{\frac{1}{2}\pi R} \frac{1}{v} dr,$$

and, since $v = c \cos \chi$, we find $T = \infty$.

7. In the system A g_{44} is constant, in B g_{44} diminishes with increasing r. Consequently in B the lines is the spectra of very distant objects must appear displaced towards the red. This displacement by the inertial field is superposed on the displacement produced by the gravitational field of the stars themselves. It is well known that the Helium-stars show a systematic displacement corresponding to a radial velocity of $+4.3$ km/sec. If we assume that about 1/8 of this is due to the gravitational field of the stars themselves,[10] then there remains for the displacement by the inertial field about 3 km/sec. We should thus have, at the average distance of the Helium stars

$$f = 1 - 2 \times 10^{-5} = \cos^2 \frac{r}{R}.$$

If for this average distance we take $r = 3 \times 10^7$ (corresponding to a parallax of $0''007$ by the formula $p = \frac{a}{r}$) this gives $R = \frac{2}{3} \times 10^{10}$. Also for the M-stars, whose average distance is probably the largest after that of the Helium-stars, Campbell[11] finds a systematic displacement of the same order. The other stars, whose average

---

[7]It might be argued that we should not see the back of the actual sun but of the sun as it was when the light left it. We could thus do without absorption, if the time taken by light to traverse the distance πR exceeded the age of the sun. With any reasonable estimale of this age, we should thus be led to still larger values of R.

[8]Contributions from the Mount WIlson Solar Observatory Nrs. 115 -117.

[9]Nature, Vol. 95, p. 701 (Aug. 26, 1915).

[10]Cf. de Sitter, *On Einstein's theory of gravitation and its astronomical consequences*, Monthly notices, Vol. 76, p. 719

[11]Lick Bulletin, Vol. 6, p. 127.

distances are smaller, also have a much smaller systematic displacement towards the red, which can very well be explained by the gravitational field of the stars themselves.

Lately some radial velocities of nebulae[12] have been observed, which are very large; of the order of 1000 km/sec. If we take 600 km/sec., and explain this as a displacement towards the red produced by the inertial field, we should, with the above value of R, find for the distance of these nebulae $r = 4 \times 10^8 = 2000$ parsecs. It is probable that the real distance is much larger.[13]

About a *systematic* displacement towards the red of the spectral lines of nebulae we can, however as yet say nothing with certainty. If in the future it should be proved that very distant objects have systematically positive apparent radial velocities, this would be an indication that the system B, and not A, would correspond to the truth. If such a systematic displacement of spectral lines should be shown not to exist, this might be interpreted either as pointing to the system A in preference to B, or as indicating a still larger value of R in the system B.

8. In the paper which has already repeatedly been quoted, Schwardschild determined the value of R for elliptical space by the condition that space should be large enough to contain the whole of our galactic system, the star-density being taken constant and equal to the value near the sun. This reasoning cannot be applied to the system A, since the field-equations give a relation between M and ρ, which contradicts Schwardschild's condition.

We have

$$\kappa \rho_0 = \frac{2}{R^2}.$$

The volume of the elliptical space is $\pi^2 R^3$. The total mass is therefore $\pi^2 R^3 \rho_0$, or

$$M = \frac{2\pi^2}{\kappa}.R.$$

If we take for M the mass of our galactic system, which can be estimated[14] at $\frac{1}{3} \times 10^{10}$ (sun = 1), then the last formula gives $R = 41$, or only about $1\frac{1}{2}$ times the distance of Neptune from the sun. This, of course, is absurd. If we use the other formula we can take for ρ_0 the star-density in the immediate neighborhood of the sun, which we estimate at 80 stars per unit of volume of Kapteyn (cube of 10 parsecs side),

---

12

N.G.C.4594	Pease	+1180	km/sec.
	Slipher	+1190	km/sec.
N.G.C.1068	Slipher	+1100	km/sec.
	Pease	+ 765	km/sec.
	Moore	+ 910	km/sec.

The nebula in Andromeda however appears to have a considerable negative velocity, viz.:

	Wright	−304	km/sec.
	Pease	−329	km/sec.
	Slipher	−300	km/sec.

[13]Eddington (Monthly Notices, Vol. 77, p. 375) estimates $r > 100000$ parsecs. This, combined with an apparent velocity of +600 km/sec., would give $R > 3 \times 10^{11}$.

[14]Communicaled by Prof. Kapteyn.

or $\rho_0 = 10^{-17}$ in astronomical units. We then find $R = 9.10^{11}$ The total mass then becomes $M = 7.10^{19}$, and consequently the galactic system would only represent an entirely negligible portion of the total world-matter.

It appears probable for many different reasons that outside our galactic system there are many more similar systems, whose mutual distances are large compared with their dimensions. If we take for the average mutual distance 10^{10} astronomical units, then an elliptical space with $R = 9.10^{11}$ could contain 7.10^6 galactic systems, of which of course only a small number are known to us by direct observation. If, however, they all actually existed, and their average mass were the same, as of our own galaxy, then their combined mass would be about 2.10^{16}, and consequently only one three-thousandth part of the world-matter would be condensed to "ordinary" matter. It is very well possible to construct a world in which the whole of the world-matter would, or at least could, be thus condensed. We must then for ρ_0 take the density not within the galactic system, but the average density over a unit of volume which is large compared with the mutual distances of the galactic systems. With the numerical data adopted above, this leads to $R = 5.10^{13}$, and there would then be more than a billion galactic systems.

All this of course is very vague and hypothetical. Observation only gives us certainty about the existence of our own galactic system, and probability about some hundreds more. All beyond this is extrapolation.

9. We now come to the case that there is gravitation, which is produced by "ordinary" matter, with the density ρ_1. I will consider the field produced by a small sphere at the origin of the system of coordinates, which I will call the "sun". Its radius is R.

In the system A the world-matter has thus everywhere the constant density ρ_0 except for values of r which are smaller than R, i. e. within the sun. There the density[15] is $\rho_0 + \rho_1$. In the system B, we have $\rho = \rho_1$, and this is zero except for $r < R$.

The line-element then has the form

$$ds^2 = -a\,dr^2 - b\left[d\psi^2 + \sin^2\psi\,d\theta^2\right] + fc^2dt^2,$$

and in a stationary state a, b, f are functions of r only. The equations become somewhat simpler if we introduce

$$l = \log a, \qquad m = \log b, \qquad n = \log f.$$

---

[15]This, of course, is not strictly in accordance with Einstein's hypothesis, by which the condensation of the world-matter in the sun should be compensated by a rarefying, or entire absence, of it elsewhere. The mass of the sun however is extremely small compared with the total mass in a unit of volume of such extent as must be taken in order to treat the density of the world-matter as constant. Therefore, if we neglect the compensation, the mass present in the unit of volume containing the sun is only *very* little in excess of that present in the other units. In the real physical world such small deviations from perfect homogeneity must always be considered as possible, and they must produce only small differences in the gravitational field.

If differential coefficients with respect to r are indicated by accents we find

$$G_{11} = m'' + \frac{1}{2}n'' + \frac{1}{2}m'(m'-l') + \frac{1}{4}n'(n'-l'),$$

$$\frac{a}{b}G_{22} = -\frac{a}{b} + \frac{1}{2}m'' + \frac{1}{4}m'(2m'+n'-l'),$$

$$-\frac{a}{f}G_{44} = \frac{1}{2}n'' + \frac{1}{4}n'(2m'+n'-l'),$$

$$G_{33} = \sin^2\psi\, G_{22}.$$

In order to write down the equations (1) we must know the values of $T_{\mu\nu}$. If all matter is at rest, and if there is no pressure or stress in it, these are: $T_{44} = g_{44}\rho$, all other $T_{\mu\nu} = 0$. These values I call $T^0_{\mu\nu}$. If we adopt these, then the equations (1) become, after a simple reduction

$$n'' + n'(m' + \frac{1}{2}n' - \frac{1}{2}l') = a(\kappa\rho - 2\lambda), \qquad (10)$$

$$m'' + \frac{1}{2}m'(m'-n'-l') = -a\kappa\rho, \qquad (11)$$

$$-\frac{a}{b} + \frac{1}{2}m'(n'+\frac{1}{2}m') = -a\lambda. \qquad (12)$$

It is easily verified that these are satisfied if we take $\rho = \rho_0$, and for $g_{\mu\nu}$ we take the values corresponding to one of the forms (4A), (4B), or (4C) of the line-element, with the conditions (3A), (3B), or (3C) respectively. Similarly for (6A), (6B) and (8A), (8B), if the accents in (10), (11), (12) denote differential coefficients with respect to r or h respectively. Consequently in the field of pure inertia we have $T_{\mu\nu} = T^0_{\mu\nu}$, i.e. by the action of inertia alone there are produced no pressures or stresses in the world-matter.

If however the mass of the sun is not neglected, then a stationary state of equilibrium, with all matter at rest, cannot exist without internal forces within this matter. The $T_{\mu\nu}$ are then different from $T^0_{\mu\nu}$. If the world-matter is considered as a continuous "fluid", then this fluid can only be at rest if there is in it a pressure or stress. If it is considered as consisting of separated material points then these cannot be at rest. The difference $T_{\mu\nu} - T^0_{\mu\nu}$ vanishes with ρ, for if $\rho = 0$, both $T_{\mu\nu}$ and $T^0_{\mu\nu}$ are zero. This difference, therefore, is of the form $\epsilon.\rho$, ϵ being of the order of the gravitation produced by the sun. The rlght-hand-members of the equations (1), and therefore also of (10), (11), (12) require corrections of the order $\kappa.\epsilon.\rho$ If these are neglected, the equations are no longer exact.

10. The mass of the sun being small, the values of a, b, f will not differ much from those of the inertial field. We can then, in the system A, and for the coordinates r, ψ, θ, put

$$a = 1 + \alpha, \qquad b = R^2 \sin^2\chi\,(1+\beta), \qquad f = 1 + \gamma,$$

and in a first approximation we can neglect the squares and products of α, β, γ. The equations then became:

$$\gamma'' + \frac{2}{R}\gamma' \cot\chi = a\kappa\rho_1, \qquad (13)$$

$$\beta'' + \frac{\cot\chi}{R}(2\beta' - \alpha' - \gamma') + \frac{2\alpha}{R^2} = -a\kappa\rho_1, \qquad (14)$$

$$\beta\,\text{cosec}^2\chi - \alpha\cot^2\chi + (\beta' + \gamma')\frac{\cot\chi}{R} = 0. \qquad (15)$$

From (13) we find, remembering that the accents denote differentiations with respect to $r = R.\chi$

$$\gamma' = \sin^2\chi = \int_0^R a\kappa\rho_1\sin^2\chi\,dr.$$

Outside the sun we have $\rho_1 = 0$. Thus if we put

$$\mathfrak{a} = R^2\int_0^R a\kappa\rho_1\sin^2\chi\,dr$$

then outside the sun

$$\gamma' = \frac{\mathfrak{a}}{R^2\sin^2\chi}$$

from which

$$\gamma = -\frac{\mathfrak{a}}{R}\cot\chi = -\frac{\mathfrak{a}}{r}. \qquad (16)$$

For $r = \frac{1}{2}\pi R$, i.e. for the largest distance which is possible in the elliptical space, we have thus $\gamma = 0$. For still larger distances, which are only possible in the spherical space, γ becomes positive, and finally for $r = \pi R$ we should have $g_{44} = \infty$, however small the mass of the sun may be, as has already been remarked above (art. 3).

If now from (14) and (15) we endeavour to determine α and β, we are met by difficulties. lt appears that the equations (13), (14), (15) are contradictory to each other. If we make the combination

$$(13) + (14) - 2 \times (15) - R\tan\chi\frac{d(15)}{dr}$$

we find

$$\gamma'\tan\chi = 0, \qquad (17)$$

which is absurd. If the equations were exact, they should, in consequence of the invariance, be dependent on each other. They are however not exact, since on the right-hand-sides terms of the order of $\epsilon.\kappa\rho$ have been neglected, ϵ being of the order of α, β, γ. In the world-matter we have[16] $\kappa\rho = \kappa\rho_0 = 2\lambda$, and these corrections can only be neglected if λ is also of the order ϵ. This has not been assumed in the equations (13), (14), (15). If we wish to assume it, then we must also develop in powers of λ. We can then use the coordinates r, ψ, θ. We put thus

$$a = 1 + \alpha, \qquad b = r^2(1 + \beta), \qquad f = 1 + \gamma.$$

---

[16] Of course, if beside the world-matter there is also "ordinary matter," i. e. if the density of the world-matter is not constant, this relation is also only approximately true, and requires a correction of the order λ, ϵ (See also art. 11).

The equations, in which now the accents denote differentiations with respect to r, then become, to the first order

$$\gamma'' + \frac{2}{r}\gamma' - \kappa\rho_1,$$

$$\beta'' + \frac{2}{r}\beta' - \frac{1}{r}(\alpha' + \gamma') = -\kappa\rho_1 - 2\lambda,$$

$$\beta - \alpha + r(\beta' + \gamma') = -\lambda r^2.$$

which are easily verified to be dependent on each other. We can thus add an arbitrary condition. If we take e.g.

$$\alpha = 2\beta$$

then we find, to the fist order, outside the sun

$$\alpha = -2\lambda r^2 + \frac{\mathfrak{a}}{r}, \qquad \beta = -\lambda r^2 + \frac{1}{2}\frac{\mathfrak{a}}{r}, \qquad \gamma = -\frac{\mathfrak{a}}{r},$$

where $\mathfrak{a} = \int_0^R \kappa\rho_1 r^2 dr$. If $\mathfrak{aa}$ is neglected these are the terms of the first order in the development of (6 A) in powers of $\lambda = \frac{1}{R^2}$.

11. Consider again the equations (10), (11), (12). If these were exact, they would be dependent on each other. They are, however, not exact, and consequently they are contradictory. If we make the combination :

$$2.\frac{d(12)}{dr} + 2[m' - l'](12) - [m' + n'].(11) - m'.(10).$$

we find[17]

$$0 = n'\, a\kappa\rho. \tag{18}$$

Consequently the equations are dependent on each other, i.e. a stationary equilibrium, all matter at rest without internal forces, is only possible, when either $\rho = 0$ or $n' = 0$, i.e. $g_{44} = $ constant. In the system A ρ is never zero, since outside the sun $\rho = \rho_0$. A stationary equilibrium is then only possible if g_{44} is constant, i.e. if no "ordinary" matter exists, for all ordinary matter will, by the mechanism of the equation (10) or (13) produce a term γ in g_{44} which is not constant. If ordinary gravitating matter does exist then not only in those portions of space which are occupied by it, but throughout the whole of the world-matter $T_{\mu\nu}$ will differ from $T^0_{\mu\nu}$. We can e.g. consider the world-matter as an adiabatic incompressible fluid. If this is supposed to be at rest, we have

$$_{22} = g_{22}p, \qquad T_{44} = g_{44}\rho_0,$$

where p is, the pressure in the world-matter. I then find

$$p = \rho_0\left(\frac{1}{\sqrt{f}} - 1\right)$$

---

[17] It is easily verified that (18) becomes identical with (17) if all terms of higher orders than the first are neglected.

and, to the first order, and for the coordinates r, ψ, θ:

$$\alpha = \beta = -\gamma = \mathfrak{a}\left(\frac{\cos 2\chi}{R \sin \chi} + \frac{1}{R}\right),$$

$$\kappa\rho_0 = 2\lambda - 3\frac{\mathfrak{a}}{R^3} = 2\lambda\left(1 - \frac{3}{2}\frac{\mathfrak{a}}{R}\right).$$

For our sun $\mathfrak{a}/R$ is of the order of 10^{-20}. For $\chi = \frac{1}{2}\pi$ we have $\gamma = 0$, and for $\chi = \pi$ we should find $\gamma = \infty$, as in the approximate solution (16), in which p was neglected.

For the planetary motion we must go to the second order. I find a motion of the perihelion amounting to

$$\delta\tilde{\omega} = -\frac{3}{2}\lambda a^2 nt. \tag{19}$$

which is of course entirely negligible on account of the smallness of λa^2. In my last paper[18] it was stated that there is no motion of the perihelion. In that paper the values $T^0_{\mu\nu}$ were used, i.e. the pressure p was neglected. The motion (19) can thus be said to be produced by the pressure of the world-matter on the planet. It will disappear if we suppose that in the immediate neighborhood of the sun the world-matter is absent.

12. In the system B outside the sun we have $\rho = 0$, and the equations are dependent on each other and can be integrated.

Within the sun $n'a\kappa\rho_1$ must be of the second order, and consequently n' must be of the first order. If we put

$$f = \cos^2\chi(1+\gamma),$$

then

$$n' = -\frac{2}{R}\tan\chi + \frac{\gamma'}{1+\gamma'}$$

thus $\tan\chi/R$ must be of the first order. Since $\chi = r/R$ we find that $1/R^2$ must be of the first order, as in system A.

Developing f in powers of $1/R$ we find, to the first order

$$f = 1 - \frac{r^2}{R^2} + \gamma.$$

In the first approximation we find for γ the same value as in the systems A and C, viz: $\gamma = -\mathfrak{a}/r$. Here however we have also the term $-r^2/R^2$. Thus classical mechanics according to Newton's law can only be used as a first approximation if this term, and consequently also $\lambda = 3/R^2$ is of the *second* order. Investigating the effect of this term on planetary motion, we find a motion of the perihelion[19]

---

[18]These Proceedings, Vol. XIX. page 1224. EDITOR's NOTE: Again, by "These Proceedings" de Sitter means KNAW, Proceedings, where this paper (On the Curvature of Space) is published.

[19]In my last paper (these Proceedings Vol. XIX, p. 1224) I found

$$\delta\tilde{\omega} = \frac{3a^3}{4\mathfrak{a}R^2}nt - \frac{cnt^2}{2R^2}.$$

amounting to

$$\delta\tilde{\omega} = \frac{3a^3}{2aR^2}nt,$$

From the condition that this shall for the earth not exceeds ay $2''$ per century we find

$$R > 10^8.$$

Then $1/R^2 < 10^{-16}$ is actually of the second order compared with $\kappa = 25.10^{-8}$ This limit of R is still considerably lower than the value which was found above from the displacement of the spectral lines. For the planetary motion – and generally for all mechanical problems which do not involve very large values of r – we can therefore in both systems A and B neglect the effect of λ entirely.

The difference is due to the use of a different system of reference, with a different time and different radius-vector, in the two cases, the formulas for the transformation of the space-variables (especially the radius-vector) from one system to the other depending on the time.

ON THE CURVATURE OF SPACE

A. A. FRIEDMANN

Translation of A. A. Фридман, О кривизне мира, *УФН* (1963) **80**, вып. 3, стр. 439-446. The paper was first published in *Журнал Русского физико-химического общества* (Zhurnal Russkogo Fiziko-Chimicheskogo Obshtestva – Journal of the Russian Physico-Chemical Association) in 1924.[1]
New translation from Russian[2] by Vesselin Petkov in: Alexander A. Friedmann, *Papers On Curved Spaces and Cosmology* (Minkowski Institute Press, Montreal 2014), pp. 11-21.

§1

1. In their known works on general cosmological questions, Einstein[3] and de Sitter[4] arrive at two possible types of the universe. Einstein obtains the so-called *cylindrical world*, in which space[5] has constant curvature, which does not change with time, wherein the radius of the curvature is associated with the total mass of matter located in space. de Sitter obtains a *spherical world*, in which not only space, but the whole world possesses, to a certain extent, a character of a world of constant curvature.[6] Both Einstein and de Sitter assume a certain form of the matter tensor, reflecting the hypothesis of disconnectedness of matter and its relative rest, in other

---

[1]EDITOR'S NOTE: The paper was first publish in German: A. Friedman, Über die Krümmung des Raumes, *Zeitschrift für Physik* **10** Nr. 1, 1922, S. 377-386. In this paper Friedmann's name is not properly transcribed; he himself wrote his name in Latin alphabet with two 'n'.

[2]EDITOR'S NOTE: There exist two English translations of Friedmann's 1922 paper (for details, see Preface of the new publication by Minkowski Institute Press above). These two English translations were done from the German publication and when compared to the original Russian text it is evident that, due to the double translation (from Russian to German, and from German to English), (i) at some places the meaning of Friedmann's explanations is not properly conveyed and (ii) often leaves the impression that Friedmann did not express himself clearly, whereas the original Russian text is clear. For this reason, Friedmann's paper was translated directly from the original Russian text.

[3]A. Einstein, Kosmologische Betrachtungen zur allgemeinen Relativitätstheorie, *Preussische Akademie der Wissenschaften, Sitzungsberichte*, 1917, 142-152.

[4]W. de Sitter, On Einstein's theory of gravitation and its astronomical consequences, *Monthly Notices Roy. Astron. Soc.*, 1916-1917.

[5]By space we will mean a space, described by a manifold of three dimensions, and will assign the term "world" to a space, described by a manifold of four dimensions.

[6]F. Klein, Ueber die Integralform der Erhaltung ersätze und die Theorie der räumlichgeschlossen Welt, *Göttinger Nach.*, 1918.

206

words, the hypothesis that the velocity of matter is sufficiently small as compared with the fundamental velocity,[7] i.e., with the velocity of light.

The purpose of the present Note is to obtain the cylindrical and spherical worlds as special cases, following from certain general assumptions, and then point out the possibility of obtaining a special world, the curvature of whose space is constant with respect to three coordinates, regarded as the space coordinates, but changes with time, i.e. depends on the fourth coordinate, regarded as the time coordinate. As far as its other properties are concerned, this new type of universe resembles Einstein's cylindrical world.

2. Our considerations are based on assumptions grouped into two classes. The first class includes assumptions, identical to those made by Einstein and de Sitter, which are related to the equations governing the gravitational potentials, and to the nature of the state and the motion of matter in space. The second class contains assumptions on the general, so to speak, geometrical nature of our world; both Einstein's cylindrical world and de Sitter's spherical world can be obtained from our hypothesis as special cases.

The assumptions of the first class are the following:

1) The gravitational potentials obey Einstein's equations with the so-called "cosmological" term, which may, in particular, be zero:

$$R_{ik} - \tfrac{1}{2}g_{ik}R + \lambda\,g_{ik} = -\kappa\,T_{ik} \quad (i,\,k = 1,2,3,4), \tag{A}$$

where g_{ik} are the gravitational potentials, T_{ik} is the matter tensor, κ is a constant, $R = g^{ik}R_{ik}$, and the tensor R_{ik} is defined by

$$R_{ik} = \frac{\partial^2 \ln\sqrt{g}}{\partial x_i \partial x_k} - \frac{\partial \ln\sqrt{g}}{\partial x_\sigma}\begin{Bmatrix} ik \\ \sigma \end{Bmatrix} - \frac{\partial}{\partial x_\sigma}\begin{Bmatrix} ik \\ \sigma \end{Bmatrix} + \begin{Bmatrix} i\alpha \\ \sigma \end{Bmatrix}\begin{Bmatrix} k\sigma \\ \alpha \end{Bmatrix}, \tag{B}$$

where x_i $(i,\,k = 1,2,3,4)$ are world coordinates, and $\begin{Bmatrix} ik \\ \sigma \end{Bmatrix}$ is the Christoffel symbol of the second kind.[8]

2) Matter is in a disconnected state and is at relative rest; or, speaking less rigorously, the relative velocities of matter are negligible compared with the velocity of light. As a result of these assumptions the matter tensor T_{ik} is defined by

$$\begin{aligned} &T_{ik} = 0, \text{ if } i \text{ and } k \text{ are not simultaneously } = 4, \\ &T_{44} = c^2\rho\,g_{44}, \end{aligned} \tag{C}$$

where ρ is the density of matter and c is the fundamental velocity; here, of course, the world coordinates are divided into two groups: x_1, x_2, x_3 called space coordinates and x_4 – the time coordinate.

3. The assumptions of the second class amount to the following:

---

[7]See this term in Eddington's book: *Espace, Temps et Gravitation*, 2ème partie, Paris, 1921 p. 10.

[8]Here the sign of R_{ik} and of the scalar curvature R is different compared to the usual sign of these quantities.

1) After separating the three space coordinates (x_1, x_2, x_3) from the four world coordinates, we will have a space of constant curvature, which can, however, change with the fourth time coordinate x_4. The interval[9] $\mathrm{d}s$, defined by $\mathrm{d}s^2 = g_{ik}\,\mathrm{d}x_i\,\mathrm{d}x_k$, can be written in the following form with the help of a corresponding change of the space coordinates:

$$\mathrm{d}s^2 = R^2(\mathrm{d}x_1^2 + \sin^2 x_1 \mathrm{d}x_2^2 + \sin^2 x_1 \sin^2 x_2 \mathrm{d}x_3^2) + 2g_{14}\mathrm{d}x_1\mathrm{d}x_4$$
$$+ 2g_{24}\mathrm{d}x_2\mathrm{d}x_4 + 2g_{34}\mathrm{d}x_3\mathrm{d}x_4 + g_{44}\mathrm{d}x_4^2$$

where R is a function only of x_4; R is proportional to the radius of the curvature of space and therefore the radius of the curvature of space may change with time.

2) In the expression for the interval $\mathrm{d}s^2$, g_{14}, g_{24}, g_{34} vanish if the time coordinate is suitably chosen, or, expressed shortly, time is orthogonal to space. It seems to me that this second assumption does not have any physical or philosophical significance and is introduced solely to simplify the calculations. It is necessary to notice that the worlds of Einstein and de Sitter are special cases of our assumptions.

The assumptions 1) and 2) enable us to write $\mathrm{d}s^2$ in the form

$$\mathrm{d}s^2 = R^2(\mathrm{d}x_1^2 + \sin^2 x_1 \mathrm{d}x_2^2 + \sin^2 x_1 \sin^2 x_2 \mathrm{d}x_3^2) + M^2\mathrm{d}x_4^2, \tag{D}$$

where R depends only on x_4, and M is, generally speaking, a function of all four world coordinates. The universe of Einstein is a special case obtained from (D) by replacing R^2 with $-R^2/c^2$ and M with 1, where R is the constant (independent of x_4 as well!) radius of the curvature of space. The universe of de Sitter is obtained when in (D) R^2 is replaced with $-R^2/c^2$ and M with $\cos x_1$:[10]

$$\mathrm{d}\tau^2 = -\frac{R^2}{c^2}(\mathrm{d}x_1^2 + \sin^2 x_1 \mathrm{d}x_2^2 + \sin^2 x_1 \sin^2 x_2 \mathrm{d}x_3^2) + \mathrm{d}x_4^2, \tag{D_1}$$

$$\mathrm{d}\tau^2 = -\frac{R^2}{c^2}(\mathrm{d}x_1^2 + \sin^2 x_1 \mathrm{d}x_2^2 + \sin^2 x_1 \sin^2 x_2 \mathrm{d}x_3^2) + \cos^2 x_1 \mathrm{d}x_4^2. \tag{D_2}$$

4. It is necessary to say a bit more on the intervals in which the world coordinates are confined; or, in other words, it is necessary to agree on which points of the manifold of four dimensions will be regarded as different. Without going into details, we will assume that the space coordinates change in the intervals: x_1 in the interval $(0, \pi)$; x_2 in the interval $(0, \pi)$; x_3 in the interval $(0, 2\pi)$. As far as the time coordinate is concerned, we will leave the question of the interval in which it changes open and will return to it below.

§2

1. Using equations (A) and (C), assuming that the gravitational potentials are defined by (D), and setting in $i = 1, 2, 3$ and $k = 4$ in equations (A), we find

$$R'(x_4)\frac{\partial M}{\partial x_1} = R'(x_4)\frac{\partial M}{\partial x_2} = R'(x_4)\frac{\partial M}{\partial x_3} = 0.$$

---

[9] See, for example, A.S. Eddington, *Espace, Temps et Gravitation*, 2ème partie, Paris, 1921.

[10] Assigning the dimension of time to the interval $\mathrm{d}s$, we denote it by $\mathrm{d}\tau$; in this case the constant κ will have the dimension of length divided by mass and in CGS units will be equal to 1.87×10^{-27}. See M. Laue, *Die Relativitätstheorie*, Bd. II, Braunschweig 1921, S. 185.

These equations lead to two cases: I) $R'(x_4) = 0$, R does not depend on x_4 and is a constant; we will call the world corresponding to this case a *stationary world*. II) $R'(x_4)$ is not zero, R depends only on x_4; we will call the world corresponding to this second case a *non-stationary world*.

Starting with the stationary world, we write down equations (A) for $i, k = 1, 2, 3$ (assuming $i \neq k$) and obtain the following system of equations:

$$\frac{\partial^2 M}{\partial x_1 \partial x_2} - \cot x_1 \frac{\partial M}{\partial x_2} = 0,$$

$$\frac{\partial^2 M}{\partial x_1 \partial x_3} - \cot x_1 \frac{\partial M}{\partial x_3} = 0,$$

$$\frac{\partial^2 M}{\partial x_2 \partial x_3} - \cot x_2 \frac{\partial M}{\partial x_3} = 0,$$

whose integration gives:

$$M = A(x_3, x_4) \sin x_1 \sin x_2 + B(x_2, x_4) \sin x_1 + C(x_1, x_4), \qquad (1)$$

where A, B, C are arbitrary functions of their arguments. Solving equations (A) for the tensor R_{ik} and eliminating the unknown density ρ^{11} from the obtained and still unused equations, and substituting the expression (1) for M in these equations, we find, after somewhat lengthy, but quite elementary calculations, that for M the following two expressions are possible:

$$M = M_0 = \text{const.} \qquad (2)$$

$$M = (A_0 x_4 + B_0) \cos x_1, \qquad (3)$$

where M_0, A_0, B_0 are constants.

In the case when M is constant, we have for the stationary world the case of the cylindrical world. In this case, it is more convenient to work with the gravitational potentials of (D$_1$); by determining the density and the quantity λ we obtain Einstein's result:

$$\lambda = \frac{c^2}{R^2}, \quad \rho = \frac{2}{\kappa R^2}, \quad M = \frac{4\pi^2}{\kappa} R,$$

where M is the total mass of the entire space.

In the other possible case, when M is determined by equation (3), we arrive, by an appropriate change of x_4[12], at de Sitter's spherical world, in which $M = \cos x_1$; using the expression (D2) we find the following relations of de Sitter's world:

$$\lambda = \frac{3c^2}{R^2}, \quad \rho = 0, \quad M = 0,$$

In this way, the stationary world can be *either Einstein's cylindrical world or de Sitter's spherical world*.

---

[11] In our case, the density ρ is an unknown function of the world coordinates x_1, x_2, x_3, x_4.
[12] This change is made with the help of the formula: $\mathrm{d}\bar{x}_4 = \sqrt{A_0 x_4 + B_0} \mathrm{d}x_4$.

2. We will now turn our attention to the study of the other possible world – the non-stationary world. In this case M is a function only of x_4. By suitably changing x_4, we can, without loss of generality, set $M = 1$. Keeping in mind the great convenience of our usual representations, we will write ds^2 in a form that is analogous to (D_1) and (D_2):

$$ds^2 = -\frac{R^2(x_4)}{c^2}(dx_1^2 + \sin^2 x_1 dx_2^2 + \sin^2 x_1 \sin^2 x_2 dx_3^2) + dx_4^2, \qquad (D_3)$$

Our task is to determine R and ρ from equations (A). It is evident that the equation (A) with differend indices do not give anything, whereas for $i = k = 1, 2, 3$ equations (A) give one relation:

$$\frac{R'^2}{R^2} + \frac{2RR''}{R^2} + \frac{c^2}{R^2} - \lambda = 0. \qquad (4)$$

With $i = k = 4$ equation (A) gives the relation:

$$\frac{3R'^2}{R^2} + \frac{3c^2}{R^2} - \lambda = \kappa\, c^2 \rho, \qquad (5)$$

where

$$R' = \frac{dR}{dx_4}, \quad R'' = \frac{d^2 R}{dx_4^2}.$$

As $R' \neq 0$, the integration of equation (4), after substituting x_4 with t for convenience, gives the equation:

$$\frac{1}{c^2}\left(\frac{dR}{dt}\right)^2 = \frac{A - R + \frac{\lambda}{3c^2}R^3}{R} \qquad (6)$$

where A is an arbitrary constant. From this equation R can be obtained by inverting an elliptic integral, i.e. by finding R from the equation

$$t = \frac{1}{c}\int_a^R \sqrt{\frac{x}{A - x + \frac{\lambda}{3c^2}x}}\, dx + B, \qquad (7)$$

where B and a are constants; here, of course, the usual change of sign of the square root should be taken care of.

Equation (5) enables us to determine ρ:

$$\rho = \frac{3A}{\kappa\, R^3}, \qquad (8)$$

where the constant A is expressed through the total mass M of space by the following relation:

$$A = \frac{\kappa\, M}{6\pi^2}. \qquad (9)$$

As the mass M is positive, it follows that A is also positive.

3. The study of the non-stationary world is based on the study of equation (6) or (7). Therein, of course, the quantity λ is not determined by itself and in our study of equation (6) or (7) we will assume that λ can take on any value. We will determine those values of the variable x which can change the sign of the square root in equation (7). Restricting ourselves to the case of a positive radius of curvature, it is sufficient to us to consider such values of x for which the quantity under the square root is zero or infinity *in the interval* $(0, \infty)$ *for* x, i.e., for positive x.

One of the values of x, for which the square root in equation (7) becomes zero, is $x = 0$; the other values of x, which can change the sign of the square root in (7), are found by studying the positive roots of the equation

$$A - x + \frac{\lambda}{3c^2} x^3 = 0$$

Denoting $\frac{\lambda}{3c^2}$ by y we can construct a family of third order curves in the (x, y)−plane

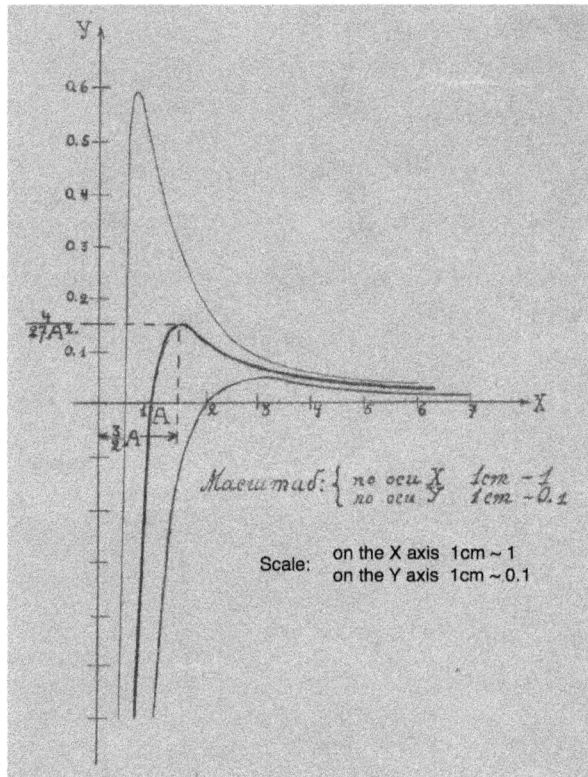

This is the original drawing in Friedmann's manuscript. Image taken from Friedmann's manuscript typed in Russian and preserved in the Ehrenfest archive.

defined by the equation:

$$y\,x^2 - x + A = 0, \tag{10}$$

where A is a parameter of the family which varies in the interval $(0, \infty)$. The curves of the family (see the figure) intersect the x-axis at the point $x = A$, $y = 0$ and have a maximum at the point:

$$x = \frac{3A}{2}, \quad y = \frac{4}{27A^2}.$$

Examining the figure shows that, for negative λ, the equation

$$A - x + \frac{\lambda}{3c^2}\, x^3$$

has one positive root x_0, lying in the interval $(0, A)$. Regarding x_0 as a function of λ and A:

$$x_0 = \theta(\lambda, A),$$

we find that θ is an increasing function of λ and an increasing function of A. Further, if λ lies in the interval $(0, 4/9(c^2/A^2))$, then our equation will have two positive roots: $x_0 = \theta(\lambda, A)$ and $x_0' = \vartheta(\lambda, A)$, wherein x_0 lies in the interval $(A, 3A/2)$ and x_0' – in the interval $(3A/2, \infty)$; $\theta(\lambda, A)$ will be an increasing function of both λ and A, whereas $\vartheta(\lambda, A)$ will be a decreasing function of λ and of A. At the end, if λ is greater than $\frac{4}{9}\frac{c^2}{A^2}$, our equation will not have positive roots at all.

Beginning our study of formula (7) we will make a remark; at the initial moment, i.e., at $t = t_0$, let the radius of the curvature be R_0. In this initial moment the square root in formula (7) will have a plus or minus sign depending on whether or not the radius of the curvature increases with time at $t = t_0$; by replacing the time t with $-t$ we can always assign a plus sign to this square root; in other words, without loss of generality, we can choose time in such a way that at the initial moment $t = t_0$ the curvature radius increases with time.

4. We consider the case when $\lambda > \frac{4}{9}\frac{c^2}{A^2}$, i.e. the case when the equation

$$A - x + \frac{\lambda}{3c^2}\, x^3$$

has no positive roots. In this case equation (7) can be written in the following way:

$$t - t_0 = \frac{1}{c}\int_{R_0}^{R}\sqrt{\frac{x}{A - x + \frac{\lambda}{3c^2}x^3}}\; \mathrm{d}x, \qquad (11)$$

where according to our remark above the square root will be always positive. It follows from here that R *will be an increasing function of* t; in this case there are no restrictions imposed on the initial value of the curvature radius R_0.

As the curvature radius could not be smaller than zero, by decreasing from R_0 with decreasing of t according to (11), the curvature radius would become zero after a certain period of time t'. Using the obvious analogy, we will call the time interval needed for the radius of curvature to increase from 0 to R_0 – the *time that elapsed since the creation of the world*;[13] this time period t' is determined from the equation:

$$t' = \frac{1}{c}\int_{0}^{R_0}\sqrt{\frac{x}{A - x + \frac{\lambda}{3c^2}x^3}}\; \mathrm{d}x, \qquad (12)$$

---

[13]The time which passed since the creation of the world is in fact the time that elapsed from the moment when space was a point ($R = 0$) to its present state ($R = R_0$); this time might be infinite.

Let us agree to call, from now on, the world we are discussing *monotonic world of the first kind.*

The elapsed time since the creation of the monotonic world of the first kind, regarded as a function of R_0, A, λ, has the following properties: 1) it increases as R_0 increases; 2) it decreases as A increases, i.e. as the mass in space increases; and 3) it decreases as λ increases. If $A > \frac{2}{3}R_0$, then for any λ the time which elapsed since the creation of the world is finite. If $A \leq \frac{2}{3}R_0$, then there can always be found a characteristic value of $\lambda = \lambda_1 = \frac{4c^2}{9A^2}$ such that as λ approaches this value, the time since the creation of the world would increase indefinitely.

5. We suppose further that λ is confined in the interval $0, \frac{4c^2}{9A^2}$; then the initial value of the curvature radius R_0 can lie in one of the three intervals: $(0, x_0)$, (x_0, x_0'), (x_0', ∞). If R_0 lies in the second interval (x_0, x_0'), then the square root in formula (7) has imaginary value and space with such initial curvature could not exist. The case when R_0 lies in the first interval $(0, x_0)$ will be considered in the next section (Section 6). Now we will consider the third case when $R_0 > x_0'$ or $R_0 > \vartheta(\lambda, A)$. In this case, consideration analogous to those in the previous section could show that R would be an increasing function of time, where R could change beginning with $x_0' = \vartheta(\lambda, A)$. The period of time from the moment when $R = x_0$ to the moment $R = R_0$ will be called the time that elapsed since the creation of the world and will denote it by t':

$$t' = \frac{1}{c} \int_{x_0'}^{R_0} \sqrt{\frac{x}{A - x + \frac{\lambda}{3c^2}x^3}} \, dx. \tag{13}$$

Let us agree to call the world we are discussing *monotonic world of the second kind.*

6. At the end we consider the case when λ is confined in the interval $(-\infty, 0)$. In this case, if $R_0 > x_0 = \theta(\lambda, A)$ the square root in (7) becomes imaginary and therefore a space with such a curvature radius could not exist. If $R_0 < x_0$, then this case will be identical to the case in the previous section, which we did not discuss there. So let us suppose that λ lies in the interval $(-\infty, \frac{4c^2}{9A^2})$ and that $R_0 < x_0$. We can show in this case, through usual considerations,[14] that R will be a periodic function of t with period $t_\text{п}$, which we call the world period and which will be determined from the expression:

$$t_\text{п} = \frac{2}{c} \int_0^{x_0} \sqrt{\frac{x}{A - x + \frac{\lambda}{3c^2}x^3}} \, dx. \tag{14}$$

where the curvature radius will change from zero to x_0. We will call such kind of world *periodic.* The period of the periodic world increases as λ increases, approaching infinity when λ approaches $\lambda_1 = \frac{4c^2}{9A^2}$. For small λ the period t_π is given by the approximation formula:

$$t_\text{п} = \frac{\pi A}{c}. \tag{15}$$

---

[14]See, for example, K. Weierstrass, Ueber eine Gattung der reel periodischer Functionen. *Mouatsber. d. Königl. Akad. d. Wissensch.*, 1866, and also J. Horn, Zur Theorie der kleinen endlichen Schwingungen. *Ztschr. f. Mathem. und Physik*, 13d. **47** 1902. The considerations of these authors should be adjusted to our case; in fact the periodicity in our case can be demonstrated through elementary considerations.

The periodic world can be viewed from two perspectives. If we assume that two events coincide, as long as the space coordinates coincide, whereas the time coordinates differ by an integer multiplied by the period, then the radius of curvature of the world increases from 0 to x_0, and after that decrease to zero. In this case the time of existence of the world will be finite.

On the other hand, if time changes from $-\infty$ to $+\infty$, i.e., if we regard two events as coincident as long as not only do their space coordinates coincide, but also their time coordinates coincide, then we arrive at a real periodicity of the space curvature.

7. The experimental evidence at our disposal is completely insufficient for carrying out numerical calculations and for finding out what kind of world is our universe. It might be that the issues of causality and of the centrifugal force could shed light on the questions discussed here. It should be noted that the "cosmological" quantity λ in our formulas is not determined; it is only an extra constant in the formulas. It might be that electrodynamical considerations could determine this quantity. Setting $\lambda = 0$ and assuming that $M = 5 \times 10^{21}$ solar masses, we obtain for the world period a quantity of the order of 10 billion years.

These figures could have, of course, merely illustrative meaning.

Petrograd,
29 May 1922

A. Friedmann,
Professor of Mechanics of the
Petrograd Polytechnic Institute

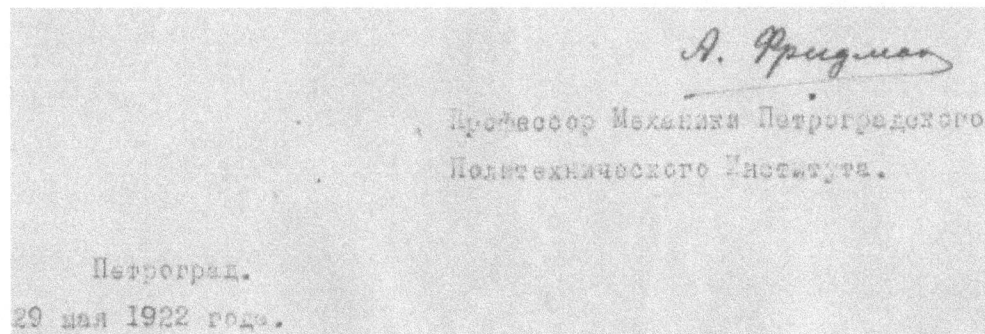

Image taken from Friedmann's manuscript typed in Russian and preserved in the Ehrenfest archive.

A Homogeneous Universe of Constant Mass and Increasing Radius accounting for the Radial Velocity of Extra-Galactic Nebulæ

G. Lemaître

Georges Lemaître, Un univers homogène de masse constante et de rayon croissant, rendant compte de la vitesse radiale des nébuleuses extra-galactiques, *Annales de la Société Scientifique de Bruxelles* 47A, pp. 49–59 (1927).

First English publication: Georges Lemaître, A homogeneous universe of constant mass and increasing radius accounting for the radial velocity of extra-galactic nebulae. *Monthly Notices of the Royal Astronomical Society* 41, 483–490 (1931). Translated from French by Georges Lemaître.[1]

1. Introduction

According to the theory of relativity, a homogeneous universe may exist such that all positions in space are completely equivalent; there is no centre of gravity. The radius of space R is constant; space is elliptic, i.e. of uniform positive curvature I/R^2; straight lines starting from a point come back to their origin after having travelled a path of length πR; the volume of space has a finite value $\pi^2 R^3$; straight lines are closed lines going through the whole space without encountering any boundary.

Two solutions have been proposed. That of de Sitter ignores the existence of matter and supposes its density equal to zero. It leads to special difficulties of interpretation which will be referred to later, but it is of extreme interest as explaining quite naturally the observed receding velocities of extra-galactic nebulae, as a simple consequence of the properties of the gravitational field without having to suppose that we are at a point of the universe distinguished by special properties.

---

[1] After some years of confusion who was the translator of the first (1931) English publication, "it is now certain that Lemaître himself translated his article" – see Jean-Pierre Luminet, Editorial note to: Georges Lemaître, "A homogeneous universe of constant mass and increasing radius accounting for the radial velocity of extra-galactic nebulae," *General Relativity and Gravitation* (2013) 45, 1619-1633, p. 1629. See also this excellent Editorial note about "An intriguing discrepancy between the original French article and its English translation" (p. 1628). A new translation of Lemaître's article by Jean-Pierre Luminet was published in the "Golden Oldies" series of *General Relativity and Gravitation* (2013) 45, 1635-1646.

The other solution is that of Einstein. It pays attention to the evident fact that the density of matter is not zero, and it leads to a relation between this density and the radius of the universe. This relation forecasted the existence of masses enormously greater than any known at the time. These have since been discovered, the distances and dimensions of extra-galactic nebulae having become known. From Einstein's formulae and recent observational data, the radius of the universe is found to be some hundred times greater than the most distant objects which can be photographed by our telescopes.

Each theory has its own advantages. One is in agreement with the observed radial velocities of nebulae, the other with the existence of matter, giving a satisfactory relation between the radius and the mass of the universe. It seems desirable to find an intermediate solution which could combine the advantages of both.

At first sight, such an intermediate solution does not appear to exist. A static gravitational field for a uniform distribution of matter without internal stress has only two solutions, that of Einstein and that of de Sitter. De Sitter's universe is empty, that of Einstein has been described as "containing as much matter as it can contain." It is remarkable that the theory can provide no mean between these two extremes.

The solution of the paradox is that de Sitter's solution does not really meet all the requirements of the problem. Space is homogeneous with constant positive curvature; space-time is also homogeneous, for all events are perfectly equivalent. But the partition of space-time into space and time disturbs the homogeneity. The coordinates used introduce a centre. A particle at rest at the centre of space describes a geodesic of the universe; a particle at rest otherwhere than at the centre does not describe a geodesic. The coordinates chosen destroy the homogeneity and produce the paradoxical results which appear at the so-called "horizon" of the centre. When we use coordinates and a corresponding partition of space and time of such a kind as to preserve the homogeneity of the universe, the field is found to be no longer static; the universe becomes of the same form as that of Einstein, with a radius no longer constant but varying with the time according to a particular law.

In order to find a solution combining the advantages of those of Einstein and de Sitter, we are led to consider an Einstein universe where the radius of space or of the universe is allowed to vary in an arbitrary way.

2. Einstein Universe of Variable Radius. Field Equations. Conservation of Energy

As in Einstein's solution, we liken the universe to a rarefied gas whose molecules are the extra-galactic nebulae. We suppose them so numerous that a volume small in comparison with the universe as a whole contains enough nebulae to allow us to speak of the density of matter. We ignore the possible influence of local condensations. Furthermore, we suppose that the nebulae are uniformly distributed so that the density does not depend on position. When the radius of the universe varies in an arbitrary way, the density, uniform in space, varies with time. Furthermore, there are generally interior stresses, which, in order to preserve the homogeneity, must reduce to a simple pressure, uniform in space and variable with time. The pressure, being two—thirds of the kinetic energy of the "molecules," is negligible with respect to the energy associated with matter; the same can be said of interior stresses in nebulae or in stars belonging to them. We are thus led to put $p = 0$.

Nevertheless it might be necessary to take into account the radiation-pressure of electromagnetic energy travelling through space; this energy is weak but it is evenly distributed through the whole of space and might afford a notable contribution to the mean energy. We shall thus keep the pressure p in the general equations as the mean radiation-pressure of light, but we shall write $p = 0$ when we discuss the application to astronomy.

We denote the density of total energy by ρ, the density of radiation energy by $3p$, and the density of the energy condensed in matter by $\delta = \rho - 3p$. We identify ρ and $-p$ with the components T_4^4 and $T_1^1 n = T_2^2 = T_3^3$ of the material energy tensor, and δ with T. Working out the contracted Riemann tensor for a universe with a line-element given by

$$ds^2 = -R^2 \, d\sigma^2 + dt^2, \tag{1}$$

where $d\sigma$ is the elementary distance in a space of radius unity, and R Is a function of time t, we find that the field equations can be written

$$3\frac{R'^2}{R^2} + \frac{3}{R^2} = \lambda + \kappa\rho \tag{2}$$

and

$$2\frac{R''}{R} + \frac{R'^2}{R^2} + \frac{1}{R^2} = \lambda - \kappa\rho. \tag{3}$$

Accents denote derivatives with respect to t. λ is the unknown cosmological constant, and κ is the Einstein constant whose value is 1.87×10^{-27} in C.G.S. units (8π in natural units).

The four identities giving the expression of the conservation of momentum and of energy reduce to

$$\frac{d\rho}{dt} + \frac{3R'}{R}(\rho + p) = 0, \tag{4}$$

which is the energy equation. This equation can replace (3). As $V = \pi^2 R^3$ it can be written

$$d(V\rho) + pdV = 0, \tag{5}$$

showing that *the variation of total energy plus the work done by radiation-pressure in the dilatation of the universe is equal to zero.*

3. Universe of Constant Mass

If $M = V\delta$ remains constant, We write, α being a constant,

$$\kappa\delta = \frac{\alpha}{R^3}. \tag{6}$$

As

$$\rho = \delta + 3p$$

we have

$$3d(pR^3) + 3pR^2 dR = 0 \tag{7}$$

and, β being a constant of integration,

$$\kappa p = \frac{\beta}{R^4} \tag{8}$$

and therefore

$$\kappa \rho = \frac{\alpha}{R^3} + \frac{3\beta}{R^4}. \tag{9}$$

By substitution in (2) we have

$$\frac{R'^2}{R^2} = \frac{\lambda}{3} - \frac{1}{R^2} + \frac{\kappa \rho}{3} = \frac{\lambda}{3} - \frac{1}{R^2} + \frac{\alpha}{3R^3} + \frac{\beta}{R^4} \tag{10}$$

and

$$t = \int \frac{dR}{\sqrt{\frac{\lambda R^2}{3} - 1 + \frac{\alpha}{3R} + \frac{\beta}{R^2}}}. \tag{11}$$

When α and β vanish, we obtain the de Sitter solution in Lanczos's form

$$R = \sqrt{\frac{3}{\lambda}} \cosh \sqrt{\frac{\lambda}{3}} (t - t_0) \tag{12}$$

The Einstein solution is found by making $\beta = 0$ and R constant. Writing $R' = R'' = 0$ in (2) and (3) we find

$$\frac{1}{R^2} = \lambda \qquad \frac{3}{R^2} = \lambda + \kappa \rho \qquad \rho = \delta$$

or

$$R = \frac{1}{\sqrt{\lambda}} \qquad \kappa \delta = \frac{2}{R^2} \tag{13}$$

and from (6)

$$\alpha = \kappa \delta R^3 = \frac{2}{\sqrt{\lambda}}. \tag{14}$$

The Einstein solution does not result from (I4) alone; it also supposes that the initial value of R' is zero. If we write

$$\lambda = \frac{1}{R_0^2}. \tag{15}$$

We have for $\beta = 0$ and $\alpha = 2R_0$

$$t = R_0 \sqrt{3} \int \frac{dR}{R - R_0} \sqrt{\frac{R}{R + 2R_0}}. \tag{16}$$

For this solution the two equations (I3) are of course no longer valid.

Writing

$$\kappa\delta = \frac{2}{R_{E^2}} \tag{17}$$

we have from (I4) and (15)

$$R^3 = R_{E^2} R_0. \tag{18}$$

The value of R_E, the radius of the universe computed from the mean density by Einstein's equation (17), has been found by Hubble to be

$$R_E = 8.5 \times 10^{28} \text{ cm} = 2.7 \times 10^{10} \text{ persec.} \tag{19}$$

We shall see later that the value of R_0 can be computed from the radial velocities of the nebulae; R can then be found from (18).

Finally, we shall show that a serious departure from (14) would lead to consequences not easily acceptable.

4. Doppler Effect due to the Variation of the Radius of the Universe

From (1) We have for a ray of light

$$\sigma_2 - \sigma_1 = \int_{t_1}^{t_2} \frac{dt}{R}, \tag{20}$$

where σ_1 and σ_2 relate to spatial coordinates. We suppose that the light is emitted at the point σ_1 and observed at σ_2. A ray of light emitted slightly later starts from σ_1 at time $t_1 + \delta t_1$ and reaches σ_2 at time $t_2 + \delta t_2$. We have therefore

$$\frac{\delta t_2}{R_2} - \frac{\delta t_1}{R_1} = 0, \qquad \frac{\delta t_2}{\delta t_1} - 1 = \frac{R_2}{R_1} - 1, \tag{21}$$

where R_1 and R_2 are the values of the radius R at the time of emission t_1 and at the time of observation t_2. If δt_1 is the period of the emitted light, δt_2 is the period of the observed light. Now δt_1 is also the period of light emitted under the same conditions in the neighbourhood of the observer, because the period of light emitted under the same physical conditions has the same value everywhere when reckoned in proper time. Therefore

$$\frac{v}{c} = \frac{\delta t_2}{\delta t_1} - 1 = \frac{R_2}{R_1} - 1 \tag{22}$$

is the apparent Doppler effect due to the variation of the radius of the universe. *It equals the ratio of the radii of the universe at the instants of observation and emission, diminished by unity.*

v is that velocity of the observer which would produce the same effect. When the light source is near enough, we have the approximate formulae

$$\frac{v}{c} = \frac{R_2 - R_1}{R_1} = \frac{dR}{R} = \frac{R'}{R} dt = \frac{R'}{R} r,$$

where r is the distance of the source. We have therefore

$$\frac{R'}{R} = \frac{v}{cr}. \tag{23}$$

From a discussion of available data, We adopt

$$\frac{R'}{R} = 0.68 \times 10^{-27} \text{ cm}^{-1} \tag{24}$$

and find from (16)

$$\frac{R'}{R} = \frac{1}{R_0\sqrt{3}}\sqrt{1 - 3y^2 + 2y^3}, \tag{25}$$

where

$$y = \frac{R_0}{R}. \tag{26}$$

Now from (18) and (26)

$$R_0^2 = R_{E^2}y^3 \tag{27}$$

and therefore

$$3\left(\frac{R'}{R}\right)^2 R_{E^2} = \frac{1 - 3y^2 + 2y^3}{y^3}. \tag{28}$$

With the adopted numerical data (24) and (19), we have

$$y = 0.0465$$

giving

$$R = R_E\sqrt{y} = 0.215R_E = 1.83 \times 10^{28}\text{cm} = 6 \times 10^9 \text{parsecs}.$$
$$R_0 = R_y = R_Ey^{\frac{3}{2}} = 8.5 \times 10^{26}\text{cm} = 2.7 \times 10^8 \text{parsecs}.$$
$$= 9 \times 10^8 \text{ light-years}.$$

Integral (16) can easily be computed. Writing

$$x^2 = \frac{R}{R + 2R_0} \tag{29}$$

it can be written

$$t = R_0\sqrt{3} \int \frac{4x^2\,dx}{(1 - x^2)(3x^2 - 1)}$$
$$= R_0\sqrt{3}\log\frac{1 + x}{1 - x} + R_0\log\frac{\sqrt{3}x - 1}{\sqrt{3}x + 1} + C. \tag{30}$$

If σ is the fraction of the radius of the universe travelled by light during time t, we have also

$$\sigma = \int \frac{dt}{R} = \sqrt{3}\int \frac{2dx}{3x^2 - 1} = \log\frac{\sqrt{3}x - 1}{\sqrt{3}x + 1} + C'. \tag{31}$$

TABLE.—*Values of σ and t.*

$\dfrac{R}{R_0}$	$\dfrac{t}{R_0}$	σ Radians	σ Degrees.	$\dfrac{v}{c}$
1	$-\infty$	$-\infty$	$-\infty$	19
2	$-4\cdot31$	$-0\cdot889$	$-51°$	9
3	$-3\cdot42$	$-0\cdot521$	-30	$5\tfrac{2}{3}$
4	$-2\cdot86$	$-0\cdot359$	-21	4
5	$-2\cdot45$	$-0\cdot266$	-15	3
10	$-1\cdot21$	$-0\cdot087$	-5	1
15	$-0\cdot50$	$-0\cdot029$	$-1\cdot7$	$\tfrac{1}{3}$
20	$0\cdot00$	$0\cdot000$	$0\cdot0$	0
25	$0\cdot39$	$0\cdot017$	1	..
∞	∞	$0\cdot087$	5	..

The following table gives values of σ and t for different values of R/R_0:

The constants of integration are adjusted to make σ and t vanish for $R/R_0 = 20$ in place of 21.5. The last column gives the Doppler effect computed from (22). The approximate formula (23) would make v/c proportional to r and thus to σ. The error is only 0.005 for $v/c = 1$. The approximate formula may therefore be used within the limits of the visible spectrum.

5. The Meaning of Equation (14)

The relation (14) between the two constants λ and α has been adopted following Einstein's solution. It is the necessary condition that the quartic under the radical in (11) may have a double root R_0 giving on integration a logarithmic term. For simple roots, integration would give a square root, corresponding to a minimum of R as in de Sitter's solution (12). This minimum would generally occur at time of the order of R_0, say 10^9 years – i.e. quite recently for stellar evolution.

If the positive roots were to become imaginary, the radius would vary from zero upwards, the variation slowing down in the neighbourhood of the modulus of the imaginary roots. In both cases the time of variation of R in the same sense Would be of the order of R_0 if the relation between λ and α were seriously different from (14).

6. Conclusion

We have found a solution such that

(1°) The mass of the universe is a constant related to the cosmological constant by Einstein's relation

$$\sqrt{\lambda} = \frac{2\pi^2}{\kappa M} = \frac{1}{R_0}.$$

(2°) The radius of the universe increases without limit from an asymptotic value R_0 for $t = -\infty$.

(3°)The receding velocities of extra-galactic nebulae are a cosmical effect of the expansion of the universe. The initial radius R_0 can be computed by formulae (24) and (25) or by the approximate formula

$$R_0 = \frac{rc}{v\sqrt{3}}.$$

This solution combines the advantages of the Einstein and de Sitter solutions.

Note that the largest part of the universe is forever out of our reach. The range of the 100-inch Mount Wilson telescope is estimated by Hubble to be 5×10^7 parsecs, or about $R/200$. The corresponding Doppler effect is 3000 km/s. For a distance of $0.087R$ it is equal to unity, and the whole visible spectrum is displaced into the infra-red. It is impossible to see ghost-images of nebulae or suns, as even if there were no absorption these images would be displaced by several octaves into the infra-red and would not be observed.

It remains to find the cause of the expansion of the universe. We have seen that the pressure of radiation does work during the expansion. This seems to suggest that the expansion has been set up by the radiation itself. In a static universe light emitted by matter travels round space, comes back to its starting-point, and accumulates indefinitely. It seems that this may be the origin of the velocity of expansion R'/R which Einstein assumed to be zero and which in our interpretation is observed as the radial velocity of extra-galactic nebulae.

REFERENCES

(1) For the different partitions of space and time in the de Sitter universe, see
K. Lanczos, *Phys. Zeits.*, **23**, 539, 1922.
H. Weyl, *Phys. Zeits.*, **24**, 230, 1923.
P. Du Val, *Phil. Mag.*, 6, **47**, 930, 1924.
G. Lemaître, *Journal of Math. and Phys.*, **4**, No. 3, May 1925.

(2) Equations of the universe of variable radius and constant mass have been fully discussed, without reference to the receding velocity of nebulae, by
A. Friedmann, "Über die Krümmung des Raümes," *Z. f. Phys.*, **10**, 377, 1922; see also
A. Einstein, *Z.f. Phys.*, **11**, 326, 1922, and **16**, 228, 1923.
The universe of variable radius has been independently studied by
R. C. Tolman, *P.N.A.S.*, **16**, 320, 1930.

(3) Discussion of the theory, and recent developments are found in
A. S. Eddington, *M. N.*, **90**, 668, 1930.

W. de Sitter, *Proc. Nat. Acad. Sci.*,**116**, 474, 1930, and *B. A. N.*, **5**, No. 185, 193, and 200 (1930).

G. Lemaître, *B. A. N.*, **5**, No. 200, 1930.

(4) Popular expositions have been given by
G. Lemaître, "La grandeur de l'espace," *Revue des questions scientifiques*, March

1929.
W. de Sitter, "The Expanding Universe," *Scientia*, Jan. 1931.

www.ingramcontent.com/pod-product-compliance
Lightning Source LLC
Chambersburg PA
CBHW051334200326
41519CB00026B/7426

* 9 7 8 1 9 2 7 7 6 3 7 0 4 *